Genetic Manipulation of the Early Mammalian Embryo

Row 1: K. Willison, D. Baltimore; U. Storb, R. Jaenisch, F. Costantini.
Row 2: P. Rigby, E. Davidson, R. Palmiter; S. Tilghman, M. Evans.
Row 3: T. Maniatis, E. Lacy; D. Hanahan, N. Copeland, N. Jenkins.
Row 4: D. Stinchcomb; L. Silver, K. Willison, T. Stewart; R. Grosschedl.

Genetic Manipulation of the Early Mammalian Embryo

Edited by

FRANK COSTANTINI
Department of Human Genetics and Development
College of Physicians and Surgeons
Columbia University

RUDOLF JAENISCH
Whitehead Institute and Department of Biology
Massachusetts Institute of Technology and
Heinrich-Pette-Institut
Hamburg, Federal Republic of Germany

COLD SPRING HARBOR LABORATORY
1985

BANBURY REPORT SERIES

Banbury Report 1: Assessing Chemical Mutagens
Banbury Report 2: Mammalian Cell Mutagenesis
Banbury Report 3: A Safe Cigarette?
Banbury Report 4: Cancer Incidence in Defined Populations
Banbury Report 5: Ethylene Dichloride: A Potential Health Risk?
Banbury Report 6: Product Labeling and Health Risks
Banbury Report 7: Gastrointestinal Cancer: Endogenous Factors
Banbury Report 8: Hormones and Breast Cancer
Banbury Report 9: Quantification of Occupational Cancer
Banbury Report 10: Patenting of Life Forms
Banbury Report 11: Environmental Factors in Human Growth and Development
Banbury Report 12: Nitrosamines and Human Cancer
Banbury Report 13: Indicators of Genotoxic Exposure
Banbury Report 14: Recombinant DNA Applications to Human Disease
Banbury Report 15: Biological Aspects of Alzheimer's Disease
Banbury Report 16: Genetic Variability in Responses to Chemical Exposure
Banbury Report 17: Coffee and Health
Banbury Report 18: Biological Mechanisms of Dioxin Action
Banbury Report 19: Risk Quantitation and Regulatory Policy

Banbury Report 20: Genetic Manipulation of the Early Mammalian Embryo

Printed in the United States of America
Cover and book design by Emily Harste

Library of Congress Cataloging in Publication Data
Main entry under title:

Genetic manipulation of the early mammalian embryo.

(Banbury report, ISSN 0198-0068 ; 20)
Includes bibliographies and indexes.
1. Mammals–Genetics. 2. Genetic engineering.
3. Developmental genetics. 4. Embryology–Mammals.
I. Costantini, Frank, 1952- II. Jaenisch, Rudolf.
III. Series.
QL738.5.G44 1985 599'.015 85-13233
ISBN 0-87969-220-0

All Cold Spring Harbor Laboratory publications may be ordered directly from Cold Spring Harbor Laboratory, Box 100, Cold Spring Harbor, New York 11724. (Phone: 1-800-843-4388) In New York State (516) 367-8425

Participants

David Baltimore, Whitehead Institute for Biomedical Research, Massachusetts Institute of Technology

Ralph L. Brinster, School of Veterinary Medicine, University of Pennsylvania

C. Thomas Caskey, Departments of Medicine and Cell Biology, Baylor College of Medicine

Verne M. Chapman, Department of Molecular Biology, Roswell Park Memorial Institute

Neal G. Copeland, Department of Microbiology and Molecular Genetics, University of Cincinnati College of Medicine

Frank Costantini, Department of Human Genetics and Development, College of Physicians and Surgeons, Columbia University

Eric H. Davidson, Division of Biology, California Institute of Technology

Martin Evans, Department of Genetics, University of Cambridge, Cambridge, England

Ronald M. Evans, Salk Institute for Biological Studies

Jeanette S. Felix, Center for Research on Mothers and Children, National Institute of Child Health and Human Development

Kirsten Fischer-Lindahl, Basel Institute for Immunology, Basel, Switzerland

Rudolf Grosschedl, Whitehead Institute for Biomedical Research, and Department of Biology, Massachusetts Institute of Technology

Douglas Hanahan, Cold Spring Harbor Laboratory

Rudolf Jaenisch, Whitehead Institute for Biomedical Research, and Department of Biology, Massachusetts Institute of Technology

Nancy A. Jenkins, Department of Microbiology and Molecular Genetics, University of Cincinnati College of Medicine

Duane Kraemer, Department of Veterinary Physiology and Pharmacology, Texas A&M University

Elizabeth Lacy, Sloan-Kettering Institute

Arnold J. Levine, Department of Molecular Biology, Princeton University

Thomas Maniatis, Department of Biochemistry and Molecular Biology, Harvard University

Frederick Naftolin, Department of Obstetrics and Gynecology, Yale University School of Medicine

John Newport, Department of Biology, University of California, San Diego

Richard D. Palmiter, Howard Hughes Medical Institute, University of Washington

Christopher Polge, ARC Institute of Animal Physiology, Animal Research Station, Cambridge, England

Peter W. J. Rigby, Department of Biochemistry, Imperial College of Science and Technology, London, England

Frank H. Ruddle, Department of Biology, Yale University

Joseph D. Schulman, The Genetics and IVF Institute

Lee M. Silver, Department of Molecular Biology, Princeton University

Davor Solter, The Wistar Institute

Timothy Stewart, Department of Genetics, Harvard Medical School

Dan T. Stinchcomb, Department of Cellular and Developmental Biology, Harvard University

Ursula Storb, Department of Microbiology and Immunology, University of Washington

Azim Surani, AFRC Institute of Animal Physiology, Cambridge, England

Shirley M. Tilghman, Institute for Cancer Research, Philadelphia

Erwin F. Wagner, European Molecular Biology Laboratory, Heidelberg, Federal Republic of Germany

Keith Willison, Chester Beatty Laboratories, Institute for Cancer Research, London, England

Preface

Some new fields may begin to move so rapidly that it becomes important to draw the various strands together to gain an overall perspective on their current status and likely future directions. Perhaps no field exemplifies this more than that comprising recent developments in the experimental manipulation of the mammalian genome. Techniques for introducing cloned genes directly into the germline, introducing genetically altered cells into early embryos, or employing developmental mutations for identifying, isolating, and then cloning the specific genes affected are collectively contributing toward an understanding of mammalian gene regulation and genetic control of development at a level not previously approachable experimentally. The practical application of these techniques, especially in farm and ranch animal husbandry, is also being developed at several centers throughout the world and they have gained considerable popular publicity.

However, it is not sheep or cows, but the mouse which remains the undisputed mammal of choice for research in this area. It is thus not surprising that the October 1984 Banbury conference from which this publication emanates was dominated by experimental work being done in that animal. Several lines of research now involve the introduction and integration into mice of cloned genes, which are retained not only through embryogenesis, but also into later generations via normal germline transmission. In some cases these genes have also been appropriately expressed and regulated in recipient animals. Experimental approaches of this sort are beginning to probe key questions concerning the site specificity of gene integration and the identification and characterization of regulatory elements controlling gene expression during development.

These collected papers of the fall 1984 Banbury conference on the genetic manipulation of the mammalian ovum and early embryo should prove of value both in gauging present capabilities as well as for gaining a perspective on the likely near future progress of this rapidly moving new field.

I am pleased to take this opportunity to acknowledge the support received for this project from the March of Dimes Birth Defects Foundation, the Fogarty International Center, and the National Institute of Child Health and Human Development. It is also once again a great pleasure to express my appreciation to Bea Toliver, the Banbury Center administrative assistant, for her usual, highly competent and personable attention to the organization of this conference and to Judith Blum, the Banbury editor, for her continued standards of excellence and assiduous attention to all aspects of production of this volume.

Michael Shodell
Director
Banbury Center

Contents

Session 1: Developmental Genetics

Molecular Studies of the Structure and Evolution of Mouse t Haplotypes

LEE M. SILVER
Department of Molecular Biology
Princeton University
Princeton, New Jersey 08544

OVERVIEW

A particular region of mouse chromosome 17 occurs in a variant form known as a *t* haplotype which is present at a high frequency in wild populations. The *t* haplotype can be considered as a genomic parasite since it is able to maintain its own integrity and it functions to ensure its own survival without providing any obvious benefit to the animal in which it resides. A series of molecular probes have been obtained that distinguish *t* haplotype genomic sequences from wild-type counterparts. These probes have been used to clarify the relationships of *t* haplotypes to each other and to the normal form of the chromosome. Additionally, these probes allow a genetic dissection and characterization of the crucial *t*-specific property of transmission ratio distortion. Finally, the accumulated data allow speculation into the origin and evolution of these unusual chromosomal variants.

INTRODUCTION

Mouse *t* haplotypes have been studied intensely by many different investigators over the 50 years since they were first discovered as naturally occurring genetic variants (Gluecksohn-Waelsch and Erickson 1970; Bennett 1975; Sherman and Wudl 1977; Silver 1981). However, it is only within the last several years that we have begun to decipher the structure of *t* haplotypes and the rationale for their existence. In fact, much of the older literature on this topic is rather misleading. What workers now call *t* haplotypes were originally known as *t* alleles or *t* mutations that were thought to be alleles of each other and dominant *T* mutations at the *T* locus. The so-called *t* mutations appeared to express pleiotropic effects on tail length, fertility, embryogenesis, male transmission ratio, and meiotic recombination.

We now realize that the simple locus model of *t* haplotypes is incorrect. There is a well-defined *T* locus near the centromere on chromosome 17, and spontaneous and induced mutations at this locus cause a shortening of tail length in heterozygotes and are lethal in homozygous embryos. However, the *t* haplotypes are not single locus mutations, and they have never been observed to occur de novo in the laboratory. *t* Haplotypes are only derived from wild mice and are recognized by their interaction with *T* mutations to produce tailless *T*/*t* animals. A large body of evidence (to be discussed below) indicates that the *t* haplotypes are structurally variant forms of a major portion of mouse chromosome 17. Within this variant chromosomal region are many normally functioning genes interspersed with a number of independent "mutant loci" that mediate the *t* effects on sperm differentiation, embryogenesis, and tail length. The *t* haplotype maintains itself as a well-defined genomic entity by suppression of recombination along its length, and it propagates itself through mouse populations by means of a male-specific transmission ratio distortion in its favor. There is no convincing evidence that the individual *t*-associated "mutant genes" bear any structural or functional homology to each other. Therefore, a *t* haplotype cannot be considered to be a gene family like the H-2 complex, but rather, a *t* haplotype must be thought of as a structural entity.

A complete *t* haplotype is operationally defined as the common *t* form that exists in wild populations of *mus domesticus* and *mus musculus*. Complete *t* haplotypes extend from a point between the centromere and *T* to a point between the H-2 complex and *Pgk-2* (Fig. 1). Recombination is suppressed along the length of *t* DNA in +/*t* heterozygotes, but normal recombination will occur along regions of *t* DNA overlap in animals heterozygous for two *t* haplotypes in *trans* configuration (Silver and Artzt 1981). The t complex is defined as the genomic region over which recombination is suppressed in heterozygotes for complete *t* haplotypes. It is important to realize that the t complex would not exist apart from its *t* haplotype-dependent definition.

The *t* haplotypes provide a system that can be used to advantage for investigations into a variety of biological fields, including chromosome evolution, meiotic recombination, population dynamics, germ cell differentiation, early embryonic development, and immunogenetics. In this report, we describe studies that have combined the tools of classical genetics with molecular probes in order to understand the structure of *t* haplotypes. The accumulated data allow speculation into the origin and evolution of these unusual genomic parasites.

RESULTS AND DISCUSSION

Over the last 5 years, a variety of molecular probes for the t complex have been developed in order to investigate the relationships among independently-derived *t*

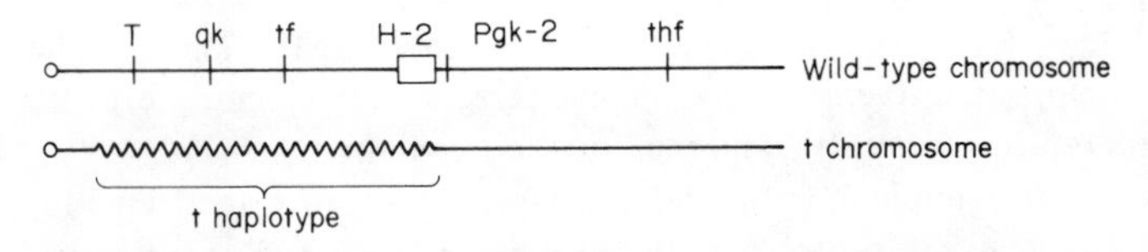

Figure 1
Schematic representation of a chromosome 17 genotype of an animal heterozygous for a complete *t* haplotype. The top line represents a wild-type form of chromosome 17 with well-characterized genetic marker loci. (*T*) Brachyury; (*qk*) quaking; (*H-2*) the *H-2* complex; (*Pgk-2*) Phophoglycerate kinase-2; and (*thf*) thin fur. The zigzag region of the bottom line represents the region of *t* DNA present in a complete *t* haplotype. A chromosome carrying a *t* haplotype is referred to as a *t* chromosome. The centromeres are on the left and the telomeres are on the right.

haplotypes and between *t* haplotypes and wild-type chromosomes. One set of probes is based upon the technique of high resolution two-dimensional gel electrophoresis. This technique has been used to identify a series of eight testicular cell polypeptides uniquely expressed in animals that carry a complete *t* haplotype (Silver et al. 1983). A comparison of testicular cell protein patterns expressed by mice carrying different *t* complex genotypes has allowed the identification of apparent allelic wild-type forms of five of these t complex proteins (TCP-1, TCP-3, TCP-4, TCP-7, and TCP-8). Cell-free translation experiments have demonstrated that the difference between the wild-type and mutant forms of each of these proteins is encoded within the mRNA sequences; this provides evidence for the location of the structural genes (e.g., *Tcp-1, Tcp-3*) for all five proteins within the t complex.

A second independent set of t complex probes was obtained through the direct cloning of random genomic sequences for microdissected fragments of the proximal portion of chromosome 17 (Roehme et al. 1984). These clones have been used to identify restriction fragment length polymorphisms (RFLPs) that are characteristic of *t* haplotypes. To date, we have identified 12 independent clones of this type that detect over 20 different regions across the length of *t* haplotype DNA (Fox et al. 1985; H. Fox et al., unpubl.). Other molecular probes for the t complex that are currently available include DNA clones of alpha globin psuedogene-4 (*Hba-ps4*) (Fox et al. 1984), the alpha crystallin I structural gene (H. Fox, unpubl.), and all of the genes located within the major histocompatibility (H-2) complex, which is an integral component of *t* haplotypes (Shin et al. 1982; Silver 1982).

A dramatic increase in our understanding of the biology of *t* haplotypes has emerged from analyses of *t* haplotypes and wild-type forms of chromosome 17 with each of the molecular probes available. One important conclusion derived from the accumulated data is that all *t* haplotypes originated from a single founder chromosome. In fact, the independent *t* haplotypes are nearly indistinguishable from each other at all coding and noncoding sequences outside of the H-2 complex (Silver et al. 1983; Fox et al. 1985). The significance of this result is emphasized by the observation that many of the DNA probes do detect polymorphisms among different wild-type chromosomes. These data suggest that the common ancestor of all currently existing *t* haplotypes occurred recently within an evolutionary timeframe.

A second conclusion derived from the accumulated data is that *t* haplotypes and wild-type chromosomes are not very different from each other at the level of primary DNA sequence (H. Fox and L.M. Silver, unpubl.). Rather, it would appear that the major structural differences between *t* and wild-type must occur at the level of gross chromosomal rearrangements. In fact, the available data provide evidence for the presence of two major non-overlapping inversions that distinguish *t* haplotypes from wild-type (Artzt et al. 1982; N. Sarvetnick et al., unpubl.). These gross rearrangements may be sufficient to explain the suppression of recombination observed between *t* and wild-type.

Although recombination is generally suppressed between *t* and wild-type, rare crossing over events do occur, and a number of the recombinant products resulting from such events have been recovered in the laboratory. These chromosomes are called partial *t* haplotypes since they retain only a portion of the original *t* haplotype DNA and express only a subset of the complete *t* haplotype properties. Partial *t* haplotypes have been used to great advantage in genetic experiments aimed at understanding the *t*-specific property of transmission ratio distortion.

Males heterozygous for a complete *t* haplotype and a wild-type chromosome can transmit their *t*-bearing chromosome to greater than 99% of their offspring. This transmission ratio distortion is crucial for the survival of *t* haplotypes in wild mouse populations. Only complete *t* haplotypes are transmitted at a high ratio.

Although mice that carry a single partial *t* haplotype cannot transmit it at a high ratio, distortion in favor of a *t* chromosome can be reconstituted in males that carry particular pairs of partial haplotypes in *cis* or *trans* configuration (Lyon 1984). With certain *trans* combinations, either the proximal, central, or distal partial *t* haplotype is transmitted at a high ratio; however, other haplotype combinations lead to near-equal ratios. Lyon and her colleagues have interpreted these data in terms of a model in which different partial *t* haplotypes carry different lengths of *t* DNA containing particular sets of "distortion loci." The basic tenet of this model is that transmission ratio distortion results from the action of a series of *t*-specific distortor loci (i.e., *Tcd-1, Tcd-2*) upon a single *t*-specific responder locus (*Tcr*). The effects of the distortor loci appear to be additive, and they can act in *cis* or *trans* to the *Tcr* locus which must be heterozygous (Fig. 2).

Molecular analysis of each of the partial *t* haplotypes used in these studies has provided strong evidence in support of the Lyon model for transmission ratio distortion (Fox et al. 1985). With additional breeding studies, it has become clear that there are at least four independent *Tcd* loci that are each correlated with a specific molecular marker (Fig. 2). The *Tcd-1* locus maps most proximally on *t* haplotypes with the T48 restriction fragment marker (Fox et al. 1985); the *Tcd-4* locus maps next with the *Tcp-1* structural gene for the p63/6.9 protein (L.M. Silver, unpubl.); the *Tcr* responder locus maps in the middle of the haplotype and is associated with the T66B restriction fragment (Fox et al. 1985); the *Tcd-3* locus maps next with several restriction fragment markers including T66C and T122; finally, the *Tcd-2* locus maps most distally with many molecular markers including T108 as well as the H-2 complex. A *t* haplotype must carry all of these loci in order to be transmitted at a high ratio. Since these loci are spread out along the entire complete *t* haplotype, it is clear that any intra-haplotype recombination event will produce recombinant chromosomes that can no longer express this property.

Although males heterozygous for a single complete *t* haplotype express a high transmission ratio, males that carry two complete *t* haplotypes are invariably and unconditionally sterile. It appears likely that this homozygous sterility is a recessive consequence of some or all of the same genes that are involved in heterozygous distortion (M.F. Lyon, pers. commun.).

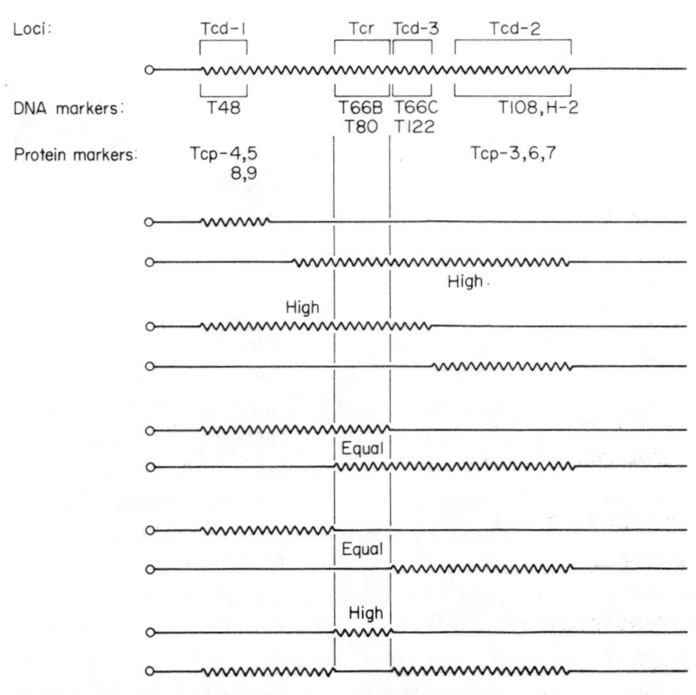

Figure 2
Genetic basis for transmission ratio distortion. The positions of three of the distorter loci (*Tcd-1, Tcd-2,* and *Tcd-3*) and the responder locus (*Tcr*) is shown at the top of the figure with associated molecular markers beneath the top chromosome representation. The *Tcd-4* locus was recently identified in association with *Tcp-1* and both are located between *Tcd-1* and *Tcr*. Four examples of genotypes with *trans* combinations of partial *t* haplotypes are shown. The chromosome transmitted at a high ratio is indicated in each case (Lyon 1984).

All complete *t* haplotypes express the same phenotypes of transmission ratio distortion, sterility, and recombination suppression. Some, but not all, express an additional phenotype of recessive embryonic lethality. To date, 15 complementing, recessive, lethal mutations have been found in association with *t* haplotypes from wild mice (Klein et al. 1984). The work of Artzt et al. (1982) clearly demonstrated that different, complementing *t* lethal mutations were nonallelic to each other. Furthermore, these studies and others (Condamine et al. 1984) show that with one exception (*tw32*), the individual lethal effects can be genetically mapped to single loci. These lethal loci are independent of the various genes involved in the other *t* haplotype properties.

With the recent advances described in this report, it has become possible to speculate on a model for the origin and evolution of *t* haplotypes that can account for nearly all of the seemingly unrelated properties characteristic of these variant genomic entities. This model is based upon the assumption that the proximal region of all forms of mouse chromosome 17 (the so-called t complex) carries genes that play important roles in haploid germ cell differentiation or sperm maturation. This assumption is supported by the observation of a large number of *t*-encoded proteins found to be expressed in the testes but not in other tissues (L.M. Silver and M. Krangel unpubl.). In addition, several non-*t* haplotype-associated genes that affect male fertility have also been localized to the t complex. Most importantly, only chromosome 17, of all the autosomes, has mutated to a form that is capable of thwarting the machinery of germ cell differentiation to its advantage.

The first event in the evolution of *t* haplotypes would be the chance accumulation on a single chromosome of a set of alleles (haplotype) at the testes-expressing loci that acted synergistically to raise their own transmission frequency to a level greater than 50%. Once such a haplotype came together, selective forces would act at the testes-expressing (distorter) loci to continue to increase the transmission ratio to even higher levels (Lewontin and Dunn 1960). However, the transmission ratio could never effectively increase above 85% since recombination events would cause 15% of all offspring to lose at least one gene required for distortion. Therefore, at this stage of evolution, chromosomal rearrangements would be selected in order to lock the various distorter loci together by recombination suppression. With continued selection for extremely high transmission ratios (of 99% or greater), *t* haplotypes would increase their frequency to significant levels in mouse populations, and males homozygous for *t* chromosomes would begin to appear. Such males are sterile (probably as a consequence of homozygousity for *t* distorter alleles) and families dominated by sterile males will not reproduce. In other words, the successfulness of these haplotypes will result in their extinction at the family level in populations. At this stage of evolution, selection will act to eliminate self-extinction in favor of *t* haplotypes which carry spontaneous recessive embryonic lethal mutations. If a *t* haplotype is lethal, all *t*-carrying animals will be heterozygous and fertile. Interestingly, over 20 years ago, Lewontin (1962) demonstrated with computer modeling that lethal *t* haplotypes were at a selective advantage relative to sterile *t* haplotypes.

The rationale for the presence of lethal mutations on *t* haplotypes has not been previously explained. Only some of the naturally occurring *t* chromosomes carry lethal mutations, so the lethal phenotype is not an obligatory feature of this system. In fact, all *t* haplotypes are virtually identical except for the presence of distinguishing lethal mutations, and at least 15 different lethal t mutations are known to exist (Klein et al. 1984). Although t effects on sperm differentiation involve interactions among multiple loci, the t lethal effects appear to be the result of simple locus mutations (with one exception described above). All of these observations are consistent with the possibility that the lethal phenotypes were acquired spontaneously during the recent evolution of *t* haplotypes. Furthermore, it appears that the t system is still in a dynamic state and that the evolution of *t* haplotypes will continue in wild mice long after we are gone.

ACKNOWLEDGMENTS

Research performed in the author's laboratory was supported by a grant from the National Institute of Child Health and Human Development.

REFERENCES

Artzt, K., P. McCormick, and D. Bennett. 1982. Gene mapping within the T/t complex of the mouse. I. t-Lethal genes are nonallelic. *Cell* **28**: 463.

Bennett, D. 1975. The T-locus of the mouse. *Cell* **6**: 441.

Condamine, H., J.-L. Guenet, and F. Jacob. 1984. Recombination between two mouse *t*-haplotypes (*tw12tf* and *tLub-1*): Segregation of lethal factors relative to centromere and tufted (*tf*) locus. *Genet. Res.* **42**: 335.

Fox, H.S., L.M. Silver, and G.R. Martin. 1984. An alpha globin pseudogene is located within the mouse t complex. *Immunogenetics* **19**: 125.

Fox, H.S., G.R. Martin, M.F. Lyon, B. Herrmann, A.-M. Frischauf, H. Lehrach, and L.M. Silver. 1985. Molecular probes define different regions of the mouse t complex. *Cell* **40**: 63.

Gluecksohn-Waelsch, S. and R.P. Erickson. 1970. The *T*-locus of the mouse: Implications for mechanisms of development. *Curr. Top. Dev. Biol.* **5**: 281.

Klein, J., P. Sipos, and F. Figueroa. 1984. Polymorphism of t-complex genes in European wild mice. *Genet. Res.* **44**: 39.

Lewontin, R.C. 1962. Interdeme selection controlling a polymorphism in the house mouse. *Am. Nat.* **887**: 65.

Lewontin, R.C. and L.C. Dunn. 1960. The evolutionary dynamics of a polymorphism in the house mouse. *Genetics* **45**: 705.

Lyon, M.F. 1984. Transmission ratio distortion in mouse *t*-haplotypes is due to multiple distorter genes acting on a responder locus. *Cell* **37**: 621.

Roehme, D., H. Fox, B. Herrmann, A.-M. Frischauf, J.-E. Edstroem, P. Mains, L.M. Silver, and H. Lehrach. 1984. Molecular clones of the mouse t complex derived from microdissected metaphase chromosomes. *Cell* **36**: 783.

Sherman, M.I. and L.R. Wudl. 1977. *Concepts in mammalian embryogenesis* (ed. M.I. Sherman), p. 136. The MIT Press, Cambridge, Massachusetts.

Shin, H.-S., J. Stavenezer, K. Artzt, and D. Bennett. 1982. Genetic structure and origin of *t* haplotypes of mice, analyzed with cDNA probes. *Cell* **29**: 969.

Silver, L.M. 1981. Genetic organization of the mouse t complex. *Cell* **27**: 239.

——— . 1982. Genomic analysis of the H-2 complex region associated with mouse *t* haplotypes. *Cell* **29**: 961.

Silver, L.M. and K. Artzt. 1981. Recombination supression of mouse *t* haplotypes due to chromatin mismatching. *Nature* **290**: 68.

Silver, L.M., J. Uman, J. Danska, and J.I. Garrels. 1983. A diversified set of testicular cell proteins specified by genes within the mouse t complex. *Cell* **35**: 35.

X Chromosome Regulation in Female Mammals

VERNE M. CHAPMAN
Department of Molecular Biology
Roswell Park Memorial Institute
Buffalo, New York 14263

OVERVIEW

The initiation of X inactivation during early embryogenesis and the subsequent maintenance of the inactive X condition throughout development are central issues in the X inactivation process. These features and the succeeding reactivation during oogenesis are unique developmental changes that produce the coordinate regulation of an entire X chromosome. This paper reviews some of the features of X chromosome expression and regulation during development. A new perspective on the process of X chromosome regulation is suggested which relates to the developmental regulation of other gene systems.

INTRODUCTION

The mechanism of X inactivation to achieve X chromosome dosage compensation is well documented in several diverse mammalian species. The coordinate regulation of an entire X chromosome occurs during early development and, once it is established, the inactive condition of an X chromosome is heritable in a clonal fashion in somatic lineages. Some of the questions we would like to address in X chromosome regulation include: (1) How is the X inactivation process initiated during development? (2) How is the inactive condition of an X chromosome maintained through successive mitotic divisions? and (3) What mechanisms are involved in reactivating inactive X chromosomes?

Understanding how the inactivation process is initiated requires some appreciation of how X chromosomes are expressed during development. Thus, the first part of this paper will review the general features of X chromosome expression and regulation during development with particular emphasis upon the early embryo.

The maintenance of the inactive X condition probably involves changes in DNA methylation and chromatin structure, but work with DNA-mediated gene transfer suggests that there may be other mechanisms operating as well. Data on the nature of inactive X chromosomes and inactive X genes may provide some insights into some of the mechanisms involved in the regulation of X chromosomes.

Reactivation of X chromosomes normally occurs in oogonia but it is difficult to achieve the reactivation of somatic cell X chromosomes experimentally using drugs and chemical agents. Recent work by Takagi et al. (1983) convincingly shows that X chromosome reactivation can be achieved in somatic cell hybrids involving embryonal carcinoma (EC) cells. These results and the earlier work by McBurney and Strutt (1980) and Martin et al. (1978) show that EC cells can have two active X chromosomes and provide new experimental opportunities to study the regulation of X chromosomes.

The work in these three areas provides the basis for reconsidering models of X chromosome regulation and the inactivation process. I would like to offer some suggestions for developing new perspectives of X chromosome regulation which draw upon the biology of X chromosome expression and the developmental regulation of other gene systems.

Finally, I would like to briefly describe work we have initiated to identify mutations in the X inactivation process which may provide a genetic approach to studying the issues of X chromosome regulation. This work involves the use of chemical mutagenesis using ethylnitrosourea (ENU). Recent work from several laboratory groups shows that it is possible to produce a frequency of 1 mutation per 2000 loci using ENU (Russell et al. 1979). This is nearly tenfold greater than the frequency produced by the highest effective dosage of X rays. The ability to produce mutations at these frequencies makes it feasible to recover from mice mutations of biological and medical interest.

DISCUSSION

X Chromosome Expression During Development

Experimental evidence that both X chromosomes are active in oocytes has been available for several years, but ascertaining the expression of X chromosomes in embryos during early cleavage stages was difficult. Cytogenetic data showed that preimplantation embryos did not have either a late labeling or asynchronously replicating X chromosomes (Takagi 1974). Thus, it was formally possible that dosage compensation is achieved during embryogenesis by either activation of a nonfunctioning X chromosome or inactivation of a transcriptionally active X chromosome. During the late 1970s several laboratories published reports showing bimodal distributions of X chromosome gene product levels among single embryos (Adler et al. 1977; Kratzer and Gartler 1978); and one report by Epstein et al. (1978) demonstrated that hypoxanthine phosphoribosyltransferase (HPRT) activity levels of XX embryos were twice those found in XY embryos. These findings were consistent with the expression of the paternally derived X chromosome in cleavage stages of development and that expression of both X chromosomes occurred in the cells of early embryos. We have recently used allelic variants of *Hprt* which differ in isoelectric point to establish qualitatively the expression of the paternal X chromosome. Our results show that the paternal *Hprt* allele is expressed in morulae stage embryos and that an intermediate charge form is present in these phenotypes which is consistent with a heteropolymer of HPRT-A/B subunits (Fig. 1). These results demonstrate that both X chromosomes are expressed during early development and that dosage compensation later in embryogenesis involves a process of inactivating a functioning X chromosome.

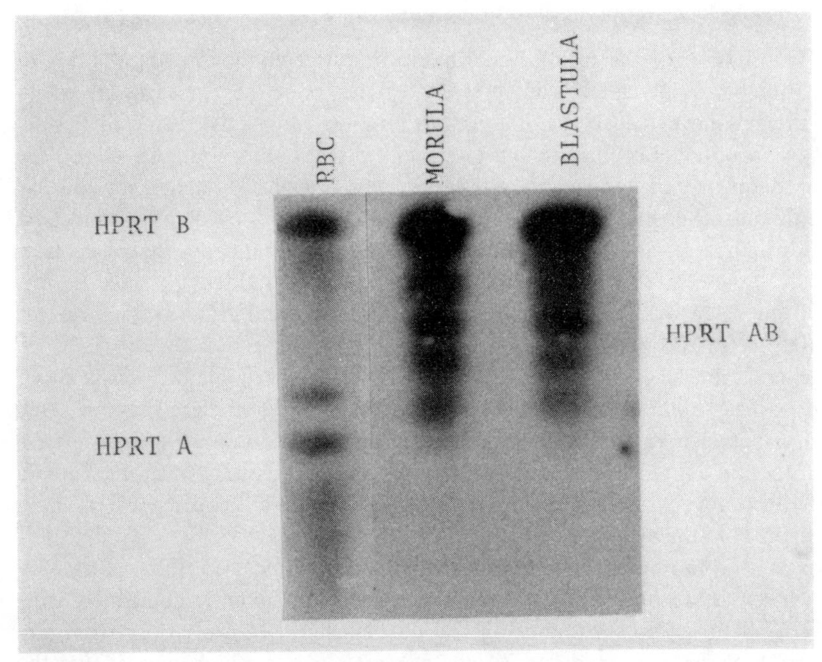

Figure 1
HPRT isoelectric focusing patterns of pooled morulae and early blastocyst stage embryos. HPRT-A and -B phenotypes are shown in lane *1* from a mixture of Hprt-B and HPRT-A RBC. The position of the HPRT-AB heteropolymers is intermediate to HPRT-A and -B homopolymer forms.

X Chromosome Inactivation in Extraembryonic Lineages

An unusual pattern of X chromosome mosaicism was observed in extraembryonic lineages of mice by Takagi and Sasaki (1975). They reported that the X chromosome from the father (*Xp*) was asynchronously replicating in extraembryonic tissues but not in the embryo proper. These findings were subsequently verified using electrophoretic variants of the PGK-1 structural gene to follow X chromosome expression (West et al. 1977). Taken together, the results of several laboratories show that only the maternal X chromosome (*Xm*) is expressed in the cell lineages derived from the trophectoderm (Frels et al. 1979) and the primitive endoderm (West et al. 1977). The experimental evidence further suggests that the preferential expression of *Xm* in these lineages is a result of nonrandom inactivation in trophectoderm and primitive endoderm. It follows that *Xm* and *Xp* must be different in early embryos and that these epigenetic differences probably arise during gametogenesis.

X Chromosome Inactivation in Embryonic Lineages

Cytodifferentiation of the trophectoderm and primitive endoderm occurs prior to implantation (Rossant and Papaioannou 1977) and, once these lineages are established, there is no evidence that they contribute to embryonic tissues of the developing conceptus. Likewise, embryonic cells do not contribute to either the trophectoderm or yolk sac endoderm lineages even though they may form yolk sac mesoderm and other placental tissues. The exact timing of X chromosome inactivation is difficult to establish in the embryonic lineages, but it probably occurs after the process takes place in trophectoderm and endoderm. Indirect evidence for this comes from early work by Gardner and Lyon (1971) and from enzyme dosage studies (Monk and Harper 1979).

In general, the inactivation of X chromosomes in embryonic lineages is either random or near random with respect to parental origin but there may be some differences between X chromosomes from different genetic sources in their relative levels of mosaic expression in both adult and embryonic tissues. Thus, the *Xm* and *Xp* differences recognized in extraembryonic lineages have either disappeared or are not recognized in the embryo.

Deviations from random mosaic expression in the mouse were first reported by Cattanach and Isaacson (1967) and they have been subsequently confirmed using the PGK-1 allelic difference (Johnston and Cattanach 1981). Deviations from random expression occur as early as 7 days of development which suggest that the difference may be a consequence of deviations from random inactivation (McMahon and Monk 1983). Cattanach has identified a putative locus responsible for these altered X-expression ratios which he has called X chromosome controlling element (Xce).

X Inactivation Associated with Cell Differentiation

The unusual patterns of nonrandom inactivation in extraembryonic lineages compared with the embryo proper and the appearance of asynchronously replicating *Xp* at the time of blastocyst development (Takagi 1978) suggest that X inactivation may accompany cell differentiation (Monk and Harper 1979). Direct evidence for this association is difficult to establish in early embryos but there is evidence from embryonal carcinoma cells to support this view. At least two different EC cell lines have been established which carry two active X chromosomes. These cell lines can be induced to differentiate with either retinoic acid (McBurney and Strutt 1980) or changes in culture conditions (Martin et al. 1978). Differentiated cells from these lines show biochemical and cytogenetic evidence of having only one active X chromosome.

Properties of Inactive X Chromosome Genes

The somatic transmission of inactive X chromosomes is a highly stable feature of the X chromosome dosage compensation mechanism. The clonal derivation of the single active X is well documented in several species and species hybrids. Attempts to select reactivated X chromosomes or derepressed X chromosome genes have not been widely successful (Kahan and DeMars 1975; Migeon et al. 1978). In general, we may conclude that inactive X chromosomes do not spontaneously reactivate at a very significant frequency and that the inactive conditions remain stable in a variety of somatic cell fusions. By contrast, 5-azacytidine treatment appears to produce a high frequency of X-chromosome gene derepression for *Hprt* (Mohandas et al. 1981). Coincident derepression of either phosphoglycerate kinase (PGK) and/or glucose-6-phosphate dehydrogenase (G6PD) also occurs but there is no direct evidence that the entire inactive X chromosome is reactivated. It should be noted that many of the reported 5-azacytidine reactivations have occurred in somatic cell hybrids or aneuploid cell lines. The significance of this is not clear but it could be a factor in the derepression events induced by this chemical agent.

The 5-azacytidine reactivation of X chromosome genes is consistent with the view that DNA methylation may play a role in the maintenance of the inactive condition. The results of DNA-mediated gene transfer experiments support this view (Liskay and Evans 1980; Chapman et al. 1982). The results of experiments reported in these papers showed that the inactive X chromosome *Hprt* gene was virtually incapable of being expressed in transformation assays. For example, in our work we observed that the inactive X chromosome allele of adult somatic cells was 50- to 100-fold less efficient in the transformation assays than the active X chromosome gene (Table 1).

We also used a similar experimental approach to examine inactive X chromosome genes in extraembryonic lineages. Namely, allelic variants of *Hprt* were used to follow the transformation of active and inactive X chromosome genes from genetically heterozygous conceptuses ($Hprt^{a/b}$) which are nonmosaic for the maternal X chromosome expression in yolk sac endoderm (Kratzer et al. 1983). We observed that frequency of transformation was greater for the active *Hprt* allelic form but that the overall efficiency of transformation of the inactive X chromosome gene was almost as high as the active X gene (Table 1). These results indicate that differential DNA methylation may be present on active and inactive X chromo-

Table 1
DNA-mediated gene transfer from heterozygous $Hprt^{a/b}$ mouse tissues into $HRPT^-$ hamster cells

Tissue phenotype	Tissue	Proportion HPRT-A transformants
B	Yolk sac endoderm	1/4 (42)[a]
A-B	Yolk sac mesoderm	1/2 (6)[a]
A-B	Lv	1/2 (25)[a]
B (TX;16)	Fetus	0 (32)[a]
	Adult Bn,Lv,Kd	1/60 (59)[b]

[a]Kratzer et al. (1983)
[b]Chapman et al. (1982)

some genes but that other mechanisms may be operating to maintain the inactive condition of the X chromosome in extraembryonic cell lineages.

We have characterized the relative extent of DNA methylation in extraembryonic lineages for centromeric and dispersed repetitive elements (Chapman et al. 1984), and we have been able to show that there is a substantial decrease in the levels of DNA methylation of elements which are normally highly methylated in somatic lineages.

This decreased level of DNA methylation is also evident in several structural genes which are either expressed or not expressed in endoderm lineages (J. Rossant, J. Sanford, and V. Chapman, in prep.). This undermethylation appears to be specific to these lineages, and work with undifferentiated and differentiated EC cells suggests that the undermethylation is a consequence of demethylation following differentiation to form endoderm (J. Sanford, J. Rossant, and V. Chapman, in prep.). The results of these experiments indicate that there is widespread demethylation of DNA in extraembryonic lineages which could account for the recovery of transformants from the inactive X chromosome *Hprt* gene which would also be undermethylated.

X Chromosome Reactivation

X chromosome reactivation is a normal feature of establishing the female germline in the ovary. X chromosome reactivation can be experimentally demonstrated by using genetic variants of an X chromosome gene such as G6PD which is a dimeric enzyme. Somatic tissues of heterozygous females express only the homopolymers of the enzyme but a heteropolymer expression can be used to monitor whether both X chromosome genes function in the same cell. Early oogonia from heterozygous female fetuses express only homopolymeric forms of G6PD but a heteropolymer is increasingly evident as germ cells are recruited into meiotic prophase (Kratzer and Chapman 1981). The results of this work and other studies indicate that only a single X chromosome is active in primordial germ cells and that X chromosome reactivation probably precedes the onset of meiotic prophase.

Experimentally induced reactivation of X chromosomes has been difficult to achieve. However, a recent report by Takagi et al. (1983) shows that reactivation may occur when somatic cells are fused with undifferentiated EC cells. These workers report that the EC cell-somatic cell hybrids do not have asynchronously replicating X chromosomes whereas these cytogenetic properties exist in other somatic cell hybrids. More convincing evidence comes from the use of EC cell fusions with either splenocytes or thymocytes from the T(X;16)16H translocation which only expresses the translocated X chromosome. These T(X;16)16H females carry the *Pgk-1*^{a} allele on the intact X chromosome which is uniformly inactive and they observed the expression of that marker in the EC cell hybrids.

Models of X Chromosome Inactivation

Several models have been developed and proposed during the past several years to describe the X inactivation process. These have been recently reviewed by Gartler and Riggs (1983). Gartler and Riggs have also developed their own model of X inactivation which involves the combined activity of X chromosome genes and autosomal gene products to induce inactivation of one X chromosome. In essence, their model involves a repressorlike mechanism which turns off an X chromosome. The primary focus of this model and other models of X inactivation has been upon the inactivation process, but the recent report by Takagi et al. (1983) suggests that it may be appropriate to consider X chromosome regulation from the perspective of reactivation as well.

I would like to propose that X chromosome expression in mammals involves two modes of regulation. First, mammalian species have evolved a mechanism which allows only one X chromosome to function in somatic cells. Second, an X chromosome reactivator mechanism is turned on in the female germ cells just prior to meiotic prophase. The reactivator mechanism overrides the somatic inactivation and reactivates the entire X chromosome.

X chromosome expression during early embryogenesis could be maintained by the continued presence of the reactivator mechanism following fertilization. The process of X chromosome inactivation could be initiated by turning off the reactivator mechanism as cells go through primary differentiation steps during embryogenesis. When the reactivator mechanism is turned off, the cell reverts to the single-active-X condition, possibly by a process of active-X selection, that is, the X chromosome which is either in the active compartment or reaches it first remains active. X chromosomes which are not in the active compartment of the nucleus are subject to enzymes which methylate DNA de novo and other secondary changes in chromatin structure. The latter could explain the recent results reported for DNA methylation of X chromosome genes which show variable sites of methylation in different lineages and even within a population of cells drawn from a single tissue (Wolf et al. 1984; Yen et al. 1984).

The nonrandom inactivation of *Xp* in extraembryonic lineages could be a result of the single-active-X property of the nucleus being maternally inherited in early embryos and of the fact that the X chromosome from the oocyte is associated with that pronucleus. These are open speculations and I have tried to avoid making a model that is tied to a specific mechanism.

The concept of an oocyte-embryo reactivator mechanism of X chromosome regulation is similar to other tissue-specific mechanisms of gene regulation. For example, the mouse genome has several families of endogenous retroviruses which show a high degree of developmentally specific regulation. Intercisternal A particle (IAP) genes share a similar pattern of developmental specificity with X chromosome regulation (Piko et al. 1984). Transcripts of IAP are present in oogenesis and during preimplantation development, but they are not expressed in postimplantation embryos. Similarly, a transposon-like retroviral element has been recently described by Brulet et al. (1983) which shows transcripts in early embryogenesis but not in later stages of development. The regulation of both of these elements is particularly relevant to the X chromosome regulation story because they are activated in early embryos or oocytes and turned off as development progresses. It is interesting to speculate that an oocyte-embryo reactivator mechanism could be acting upon X chromosome-specific enhancer elements as well.

Genetic Approaches to Studying X Chromosome Regulation

Our primary interest is to identify and characterize the genetic elements involved in X chromosome regulation. One approach we have initiated during this past year is to ask whether we can induce mutations of the X inactivation process using the chemical mutagen ENU. We are hopeful that we can produce mutations which will be informative about X chromosome regulation, but we also expect that we can induce other mutations of biological and medical interest as well.

Several laboratories have initiated work with ENU to produce specific mouse mutations for α- and β-globin genes (Popp et al. 1983), lethal and morphological mutants of the t complex on chromosome 17 (Bode 1984), and mouse models of phenylketonuria (V. Bode, pers. comm.). In our own laboratory we are producing mutations of lysosomal enzyme genes and surveying for mutants of *Hprt, Pgk-1,* G6PD, and X-linked muscular dystrophy. Mutations for some of these genes have been found and the prospect of finding more will increase as our experience increases in using this new tool for mammalian genetic research.

REFERENCES

Adler, D.A., J.D. West, and V.M. Chapman. 1977. Expression of alpha-galactosidase in preimplantation mouse embryos. *Nature* **267**: 838.

Bode, V.C. 1984. Ethylnitrosourea mutagenesis and the isolation of mutant alleles for specific genes located in the t-region of mouse chromosome 17. *Genetics* **108**: 457.

Brulet, P., M. Kaghad, Y.-S. Xu, O. Croissant, and F. Jacob. 1983. Early differential tissue expression of transposon-like repetitive DNA sequences of the mouse. *Proc. Natl. Acad. Sci. USA* **80**: 5641.

Cattanach, B.M. and J.H. Isaacson. 1967. Controlling elements in the mouse X chromosome. *Genetics* **57**: 331.

Chapman, V., L. Forrester, J. Sanford, N. Hastie, and J. Rossant. 1984. Cell lineage-specific undermethylation of mouse repetitive DNA. *Nature* **307**: 284.

Chapman, V.M., P.G. Kratzer, L.D. Siracusa, B.A. Quarantillo, R. Evans, and R.M. Liskay. 1982. Evidence for DNA modification in the maintenance of X-chromosome inactivation of adult mouse tissues. *Proc. Natl. Acad. Sci. USA* **79**: 5357.

Epstein, C.J., S. Smith, B. Travis, and G. Tucker. 1978. Both X chromosomes function before visible X chromosome inactivation in female mouse embryos. *Nature* **274**: 500.

Frels, W.I., J. Rossant, and V.M. Chapman. 1979. Maternal X chromosome expression in mouse chorionic ectoderm. *Dev. Genet.* **1**: 123.

Gardner, R.L. and M.F. Lyon. 1971. X chromosome inactivation studied by injection of a single cell into the mouse blastocyst. *Nature* **231**: 385.

Gartler, S.M. and Riggs, A.D. 1983. Mammalian X-chromosome inactivation. *Annu. Rev. Genet.* **17**: 155.

Johnston, P.G. and B.M. Cattanach. 1981. Controlling elements in the mouse. IV. Evidence of non-random X-inactivation. *Genet. Res.* **37**: 151.

Kahan, B. and R. DeMars. 1975. Localized depression on the human inactive X chromosome in mouse-human cell hybrids. *Proc. Natl. Acad. Sci. USA* **72**: 1510.

Kratzer, P.G. and V.M. Chapman. 1981. X chromosome reactivation in oocytes of *Mus caroli. Proc. Natl. Acad. Sci. USA* **78**: 3093.

Kratzer, P.G. and S.M. Gartler. 1978. HGPRT activity changes in preimplantation mouse embryos. *Nature* **274**: 503.

Kratzer, P.G., V.M. Chapman, H. Lambert, R. Evans, and R.M. Liskay. 1983. Differences in the DNA of the inactive X chromosome of fetal and extraembryonic tissues of mice. *Cell* **33**: 37.

Liskay, R.M. and R.J. Evans. 1980. Inactive X chromosome DNA does not function in DNA-mediated cell transformation for the hypoxanthine phosphoribosyltransferase gene. *Proc. Natl. Acad. Sci. USA* **77**: 4895.

Martin, G.R., C.J. Epstein, B. Travis, G. Tucker, S. Yatziv, D.W. Martin, Jr., S. Clift, and S. Cohen. 1978. X-chromosome inactivation during differentiation of female teratocarcinoma stem cells *in vitro. Nature* **271**: 329.

McBurney, M.W. and B.J. Strutt. 1980. Genetic activity of X chromosomes in pluripotent female teratocarcinoma cells and their differentiated progeny. *Cell* **21**: 357.

McMahon, A. and M. Monk. 1983. X-chromosome activity in female mouse embryos heterozygous for Pgk-1 and Searle's translocation, T(X;16)16H. *Genet. Res.* **41**: 69.

Migeon, B.R., J.A. Sprenkle, and T.D. Tai. 1978. Studies of human-mouse cell hybrids with respect to X-chromosome inactivation. In *Genetic mosaics and chimeras in mammals* (ed. L.B. Russell), p. 329, Plenum Press, New York.

Mohandas, T., R.S. Sparkes, and L.J. Shapiro. 1981. Reactivation of an inactive human X chromosome: Evidence for X inactivation by DNA methylation. *Science* **211**: 393.

Monk, M. and M.I. Harper. 1979. Sequential X chromosome inactivation coupled with cellular differentiation in early mouse embryos. *Nature* **281**: 311.

Piko, L., M.D. Hammons, and K.D. Taylor. 1984. Amounts, synthesis, and some properties of intracisternal A particle-related RNA in early mouse embryos. *Proc. Natl. Acad. Sci. USA* **81**: 488.

Popp, R.A., E.G. Bailiff, L.C. Skow, F.M. Johnson, and S.E. Lewis. 1983. Analysis of a mouse α-globin gene mutation induced by ethylnitrosourea. *Genetics* **105**: 157.

Rossant, J. and V.E. Papaioannou. 1977. The biology of embryogenesis. In *Concepts in mammalian embryogenesis* (ed. M.I. Sherman), p. 1, MIT Press, Cambridge, Massachusetts.

Russell, W.L., E.M. Kelly, P.R. Hunsicker, J.W. Bangham, S.C. Maddux, and E.L. Phipps. 1979. Specific-locus test shows ethylnitrosourea to be the most potent mutagen in the mouse. *Proc. Natl. Acad. Sci. USA* **76**: 5818.

Takagi, N. 1974. Differentiation of X chromosomes in early female mouse embryos. *Exp. Cell Res.* **86**: 127.

——— . 1978. Preferential inactivation of the paternally derived X chromosome in mice. In *Genetic mosaics and chimeras in mammals* (ed. L.B. Russell), p. 341, Plenum Press, New York.

Takagi, N. and M. Sasaki. 1975. Preferential inactivation of the paternally derived X chromosome in the extraembryonic membranes of the mouse. *Nature* **256**: 640.

Takagi, N., M.A. Yoshida, O. Sugawara, and M. Sasaki. 1983. Reversal of X-inactivation in female mouse somatic cells hybridized with murine teratocarcinoma stem cells in vitro. *Cell* **34**: 1053.

West, J.D., W.I. Frels, V.M. Chapman, and V. Papaioannou. 1977. Preferential expression of the maternally derived X-chromosome in the mouse yolk sac. *Cell* **12**: 873.

Wolf, S.F., D.J. Jolly, D.K. Lunnen, T. Friedmann, and B.R. Migeon. 1984. Methylation of the hypoxanthine phosphoribosyltransferase locus on the human X chromosome: Implications for X-chromosome inactivation. *Proc. Natl. Acad. Sci. USA* **81**: 2806.

Yen, P.H., P. Patel, A.C. Chinault, T. Mohandas, and L.J. Shapiro. 1984. Differential methylation of hypoxanthine phosphoribosyltransferase genes on active and inactive human X chromosomes. *Proc. Natl. Acad. Sci. USA* **81**: 1759.

Tissue-specific Expression of Cloned ∝-Fetoprotein Genes in Teratocarcinoma Cells and Mice

SHIRLEY M. TILGHMAN,* RICHARD W. SCOTT,* THOMAS F. VOGT,* ROBB KRUMLAUF,* ROBERT E. HAMMER,† AND RALPH BRINSTER†
*Institute for Cancer Research
Philadelphia, Pennsylvania 19111
†University of Pennsylvania School of Veterinary Medicine
Philadelphia, Pennsylvania 19104

OVERVIEW

The first tissue to express α-fetoprotein (AFP) during mouse embryogenesis is the visceral endoderm of the developing yolk sac. To identify *cis*-acting DNA elements involved in the activation of this gene during differentiation, we have introduced, via DNA-mediated transformation, modified copies of the mouse AFP gene into murine F9 embryonal carcinoma cells. These cells can be stimulated to differentiate in culture to cell types very similar to the parietal and visceral endoderms of the developing yolk sac. In five of twelve F9 transformants carrying exogenous copies of the mouse AFP gene, the differentiation to either parietal or visceral endoderm is accompanied by tissue-specific induction of the exogenous template in a manner qualitatively, but not quantitatively, identical to that of the resident AFP gene. Identical constructs have also been introduced into the germline of mice. The exogenous AFP gene is expressed in a tissue-specific manner in fetal transgenic animals, and is developmentally regulated in neonatal liver.

INTRODUCTION

In the mouse the development of the extra-embryonic yolk sac begins early in embryogenesis. Toward the end of the fourth day after fertilization the progenitor tissue, primary endoderm, segregates from the inner cell mass (Gardner 1982; Nadijcka and Hillman 1974). By day 7, the morphologically distinct parietal and visceral endodermal layers are evident (Solter and Damjanov 1973) and constitute the distal and proximal cell walls of the yolk sac, respectively. Parietal endoderm cells rest on a thick basement membrane adjacent to the trophectoderm whereas visceral endoderm cells are associated with a thin basement membrane that separates them from the inner cell mass.

The study of the formation of parietal and visceral endoderm during early mouse embryogenesis has been greatly facilitated by the establishment of embryonal carcinoma (EC) cell lines derived from the stem cells of teratocarcinomas (Kleinsmith and Pierce 1964). EC cells have many properties of the ectodermal cells of the inner cell mass (Martin 1975) and cell lines exist that are capable of differentiating in culture to a variety of tissue types including extra-embryonic endoderm. One such line, F9, has been well characterized and has a very low frequency of spontaneous differentiation (Bernstine et al. 1973). Strickland and coworkers (Strickland and Mahdavi 1978; Strickland et al. 1980) have shown that after treatment with retinoic acid and cAMP a monolayer of F9 EC cells will differentiate to a cell closely resembling parietal endoderm. If F9 EC cells are treated instead as aggregates with retinoic acid, a majority of the cells on the outer surface of the aggregate acquire a phenotype very similar to that of visceral endoderm and consequently synthesize α-fetoprotein (AFP) (Hogan et al. 1981), a marker specific for visceral endoderm.

To understand the mechanisms underlying gene activation during early mouse development we have chosen to use the F9 system to study the activation of the AFP gene in visceral endoderm. The conclusions we reach are being corroborated by parallel experiments in which the expression of cloned AFP genes introduced into the germline of mice is being examined.

RESULTS

Transfection of F9 Cells with an AFP Minigene

In order to identify any *cis*-acting DNA elements within the AFP genomic locus that are required for the activation of the AFP gene upon differentiation in culture, modified copies of the mouse AFP gene were introduced into F9 EC cells. The resulting cell lines were examined to determine whether the pattern of expression of the AFP gene in the F9 stem and differentiated cells could be reproduced from the exogenous AFP templates (Scott et al. 1984).

Stable transformants were established by cotransformation of an F9 TK^- EC cell line with unlinked plasmids containing the 3.4 kb *Bam*HI fragment of the herpes thymidine kinase (TK) gene (Enquist et al. 1979) and an AFP minigene termed YZE (Fig. 1). YZE is a 5 coding block gene containing 14 kilobase pairs (kb) of 5′ flanking DNA, the first 3 and the last 2 coding blocks of the AFP gene and 0.4 kb of 3′ flanking DNA. This construction contains all of the DNA between the 3′ terminus of the upstream albumin gene and the 5′ terminus of the AFP gene, and thus will test whether putative *cis*-acting DNA elements required for proper AFP gene regulation are located in this intergenic region. The mature transcript from the YZE minigene should be approximately 600 bases in length and therefore distinguishable from the endogenous AFP mRNA, approximately 2.2 kb in length. Eleven clonally derived transformants were chosen that contained varying copy numbers of YZE DNA, and at least 1 kb of 5′ flanking DNA intact.

Tissue Specificity of YZE mRNA Expression

We next determined the qualitative and quantitative pattern of expression from the YZE templates, relative to the endogenous gene, in the undifferentiated and differ-

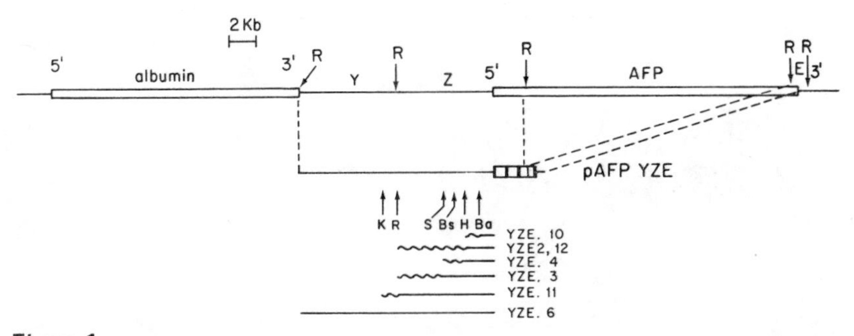

Figure 1
Structure of pAFPYZE and its configuration in several single copy F9 transformants. The minigene pAFPYZE was constructed by ligating in two stages the 1.1-kb *Eco*RI E fragment that encompasses the last two coding blocks of the AFP gene along with 450 bp of 3′ flanking DNA, the 9.8-kb *Eco*RI Z fragment that encompasses the first three coding blocks of the AFP gene along with 7.4 kb of 5′ flanking DNA, and the 7.0-kb *Eco*RI Y fragment (Ingram et al. 1981; Kioussis et al. 1981).

The line drawings (*below*) illustrate the approximate breakpoints in the 5′ flanking regions of the integrated YZE DNA in the low copy transformants. The straight line represents sequences contained in the transformants, and the wavy line represents the region where the breakpoint occurs. The enzyme digests used to map these sites are illustrated as Ba, *Bam*HI; H, *Hin*dIII; Bs, *Bst*EII; S, *Sma*I, E, *Eco*RI; K, *Kpn*I. (Reprinted, with permission, from Scott et al. 1984.)

entiated cultures of each transformant. Total poly A^+ RNA was isolated from undifferentiated EC cell monolayers and 15-day visceral endoderm cultures of each transformant. The RNA was electrophoresed on denaturing agarose gels, transferred to nitrocellulose, and hybridized with a first AFP coding block probe. The results for eight of the transformants appear in Figure 2A.

Authentic AFP mRNA was detected only in the visceral endoderm RNA (*V* lanes) for each transformant, confirming the presence of visceral endoderm in the cell aggregates. RNA species of the appropriate size for YZE mRNA were detected at relatively low levels in YZE.7,8,9,1 and 5 visceral endoderm RNA but absent in the corresponding undifferentiated EC cells. As for YZE.12 and 6 and four YZE transformants not shown in the figure, no significant amounts of RNA species corresponding to the predicted size of YZE mRNA were detected in either the EC cells or visceral endoderm. As expected, an AFP cDNA probe specific for internal coding blocks not represented in the YZE minigene hybridized with the endogenous AFP mRNA but not with the 600-base RNA (Fig. 2B).

The tissue specificity of expression from the exogenous YZE templates was investigated by asking whether YZE mRNA was present in parietal endoderm derived from the YZE EC cells. Endoderm cultures were established by treating EC cell monolayers with retinoic acid and cAMP. When RNA extracted from parietal cultures of these transformants was analyzed by gel electrophoresis and hybridiza-

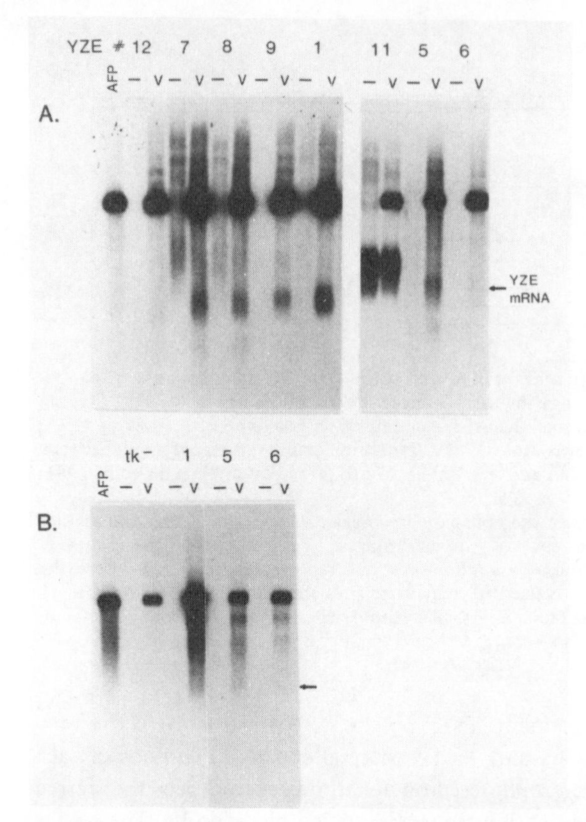

Figure 2
Analysis of AFP and YZE mRNAs in EC cell and visceral endoderm cultures of YZE transformants. Total poly A^+ RNA from the EC cells (– lanes) and 15-day visceral endoderm (*V* lanes) cultures were electrophoresed in a 6% formaldehyde, 1.5% agarose gel; and the RNA was transferred to nitrocellulose. (*Panel A*) The nitrocellulose filters were hybridized to a labeled probe that spans the first AFP coding block. (*Panel B*) The nitrocellulose filters were hybridized with the labeled *Pst*I insert of the cDNA clone, pAFP2 (Tilghman et al. 1979). This sequence contains AFP coding blocks 7-13. The AFP lanes in panels A and B correspond to 20 ng of poly A^+ fetal mouse liver RNA. (Reprinted, with permission, from Scott et al. 1984.)

tion to a first coding block probe, neither the endogenous AFP nor YZE mRNAs could be detected (Scott et al. 1984). From this we conclude that the qualitative pattern of expression from the endogenous AFP gene and the exogenous YZE genes is identical in the three cell types examined.

The one exception to this finding was YZE.11, where a transcript larger than expected for YZE mRNA was expressed constitutively. By SI mapping, it was determined that this RNA was initiated at a site upstream of the AFP cap site and that the read-through transcripts were being spliced at a cryptic splice site 33 bp upstream of the AFP cap site (data not shown).

Temporal Expression of YZE mRNA

It has been suggested that the production of visceral endoderm in the outer layer of aggregating F9 cells occurs as an orderly progression of multiple events (Grover et al. 1983a) and maximum expression of AFP requires the production of an organized epithelium properly aligned on a thin basement membrane (Grover et al. 1983b). The very low levels of YZE mRNA in all the positive transformants raised the issue of whether one or more of the stages in AFP mRNA induction was not occurring properly. For this reason, we examined the time course for induction of YZE mRNA during visceral endoderm formation in two transformants. As shown in Figure 3, AFP mRNA was detectable at very low levels by day 5 in both transformants, increased dramatically (100–500-fold) between day 7 and 9, and continued to increase through day 15. YZE mRNA was barely detectable by day 7 in

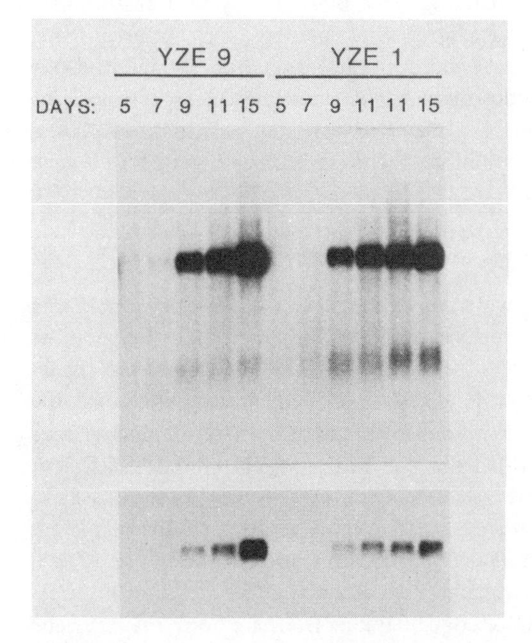

Figure 3
Time course of AFP mRNA and YZE mRNA production during visceral endoderm formation. Embryoid body cultures of YZE.9 and YZE.1 transformants were harvested at the indicated days after seeding in bacteriological plates in the presence of retinoic acid, and RNA was extracted. For each sample, 7 μg of poly A^+ RNA was electrophoresed on a denaturing agarose gel, transferred to nitrocellulose filters, and hybridized to a labeled probe that spans the first AFP coding block. The day 5 through day 15 samples for YZE.9 were collected from one series of embryoid body cultures. For YZE.1, the day 5 through the first day 11 samples were collected from one series of embryoid body cultures, and the second day 11 and day 15 samples were harvested from a second series of embryoid body cultures. The top panel is a 48-hour exposure, and the bottom panel a 3-hour exposure of the same nitrocellulose filter. (Reprinted, with permission, from Scott et al. 1984.)

both transformants; however these levels increased only two to threefold by day 15. These results suggest that YZE mRNA can be expressed in a tissue-specific fashion but the signals responsible for the further amplification of AFP mRNA that is initiated between day 7 and day 9 of aggregation is deficient in the YZE minigene construct.

Gene Copy Number and YZE Inducibility

When the configuration of DNA in the eleven transformants was determined by Southern gel analysis, it was noted that the five transformants capable of inducing YZE mRNA were all multiple copy DNA integrants (YZE.1,5,7,8,9). Only one of the noninducers, YZE.6, contained multiple inserts. This correlation may be the result of a simple dosage effect. However, it is equally possible that the single copy integrants have not retained sufficient colinear DNA in the critical region to confer inducibility. Thus, the extent of 5′ flanking DNA in each of these transformants was examined by restriction enzyme digestions. The low copy transformants YZE.2,3,4 and 12 had intact copies of the structural gene, but 5′ flanking DNA breakpoints located between map positions -7400 and -2000, respectively (see Figure 1). The 5′ junction site for YZE.11 lay between map positions -8400 and -7400, respectively. Thus the only noninducer with an intact integrated YZE gene was YZE.6.

The Regulation of AFP Minigenes in the Germline of Mice

To confirm and extend the results we have obtained in the F9 teratocarcinoma cells, we have also introduced similar AFP minigene constructs into the germline of mice, via ooctye injection (Krumlauf et al. 1985). The analysis of the tissue-specific expression of the minigenes in a variety of fetal tissues, including the visceral endoderm of the yolk sac, the fetal liver, and gut where the endogenous AFP gene is also expressed, has revealed that the YZE minigene contains sufficient DNA sequence information to confer tissue-specific expression. These mice have allowed us to now ask whether such a construct can be developmentally regulated. Transcription of the AFP gene is shut off shortly after birth in the mouse (Tilghman and Belayew 1982). When transgenic mice carrying either single or multiple copies of an AFP minigene that has 7 kb of 5′ flanking DNA are born, the minigene transcript decreases in liver at the same time and to a similar extent as the endogenous mRNA. Thus we can conclude that the DNA signals necessary for developmental regulation are also contained within the DNA that has been introduced into the mice. Experiments are continuing to identify more precisely the location of these sequences.

DISCUSSION

The F9 EC cell is a multipotent stem cell capable of differentiation into two different endoderm cells as well as into adrenergic neuron-like cells (Liesi et al. 1983). Before differentiation the AFP gene is completely silent in these cells, and it is activated selectively upon formation of visceral endoderm (Hogan et al. 1981). The objective of this study was to establish stable transformants of an EC cell line that had incorporated copies of a modified AFP minigene and then determine if expression from the endogenous AFP gene and the exogenous integrants could be activated in concert. This required that the AFP minigene promoter be inactive in the EC cell, a condition that held in all the transformants examined.

For five of the YZE transformants, the regulation of YZE mRNA expression during differentiation is qualitatively identical to authentic AFP mRNA in that it was induced selectively in visceral endoderm. No transformants were isolated in which the expression of authentic YZE mRNA was either constitutive in stem cells or induced in parietal endoderm. Thus one must conclude that the DNA in YZE is sufficient to confer tissue specificity. This conclusion is reinforced by the fact that the expression of the minigene is also tissue-specific when placed in the germline of mice.

The major difference between the activation of the endogenous AFP and exogenous YZE mRNAs was the relative low levels of YZE mRNA. The data from the time course for induction of AFP and YZE mRNAs showed that while both mRNAs were detected by day 7 of aggregation, after day 7, the levels of AFP mRNA rose dramatically while YZE mRNA increased only two to threefold. It has been argued that there are two stages to the activation of AFP gene expression, an initial activation event that occurs during the organization of the visceral endoderm and a secondary induction that occurs after formation of the basement membrane (Grover et al. 1983a,b). The mechanism underlying the second increase in AFP mRNA after day 7 has not been elucidated. It may occur via a selective stabilization of AFP mRNA, which is known to have a very long half-life (60–120 hours) in visceral yolk sac (Andrews et al. 1982). In that case, the failure of YZE mRNA to accumulate could be the result of the absence of internal sequences necessary for that stability. On the other hand, the second increase in AFP mRNA from day 9 to day 15 could occur via an increase in transcription rate, in which case the limited induction of YZE mRNA could be explained by an inefficient transcriptional induction. In that case, one must conclude that the YZE minigene is lacking sequences for optimal transcription of the gene, or that its integration into an inappropriate chromosomal locus has a repressing effect.

There is an interesting correlation between multiple integrated copies of YZE DNA and the inducibility of YZE mRNA. There are at least three possible explanations for this correlation. First, it may be a dosage effect. In order to detect YZE mRNA, there must be multiple copies of the template to compensate for very inefficient expression of the exogenous template. However, there does not seem to be a firm correlation with copy number because the transformant that accumulates the highest level of YZE mRNA, YZE.1, contains the least number of YZE DNA copies among the multiple copy transformants. Second, as has been frequently observed in other systems, a large proportion of the YZE DNA in the multiple copy transformants appears to be integrated in tandem arrays. These may serve to isolate the internal copies of the YZE templates from any repressing activity from the surrounding chromosomal DNA. A third explanation may be related to the fact that none of the low copy transformants has integrated an intact copy of the 14 kb of 5′ flanking YZE DNA whereas at least several intact copies are present in each multiple copy transformant (see Figure 1). Therefore, the possibility remains that an essential element in the 5′ flanking DNA required for activation has been deleted in all the low copy transformants. It will be important to generate full length single copy transformants to distinguish between these models.

ACKNOWLEDGMENTS

This work was supported by Grants CA06927, CA28050 and RR05539 from the National Institutes of Health and an appropriation from the Commonwealth of Pennsylvania, also from the National Institutes of Health Training Grant CA09035.

REFERENCES

Andrews, G.K., R.G. Janzen, and T. Tamaoki. 1982. Stability of α-fetoprotein messenger RNA in mouse yolk sac. *Dev. Biol.* **89**: 111.

Bernstine, E.G., M.L. Hooper, S. Grandchamp, and B. Ephrussi. 1973. Alkaline phophatase activity in mouse teratoma. *Proc. Natl. Acad. Sci. USA* **70**: 3899.

Enquist, L.W., G.F. Vande Woude, M. Wagner, J.R. Smiley, and W.C. Summers. 1979. Construction and characterization of a recombinant plasmid encoding the gene for the thymidine kinase of *Herpes simplex* type 1 virus. *Gene* **7**: 335.

Gardner, R.L. 1982. Investigation of cell lineage and differentiation in the extraembryonic endoderm of the mouse embryo. *J. Embryol. Exp. Morphol.* **68**: 175.

Grover, A., R.G. Oshima, and E.D. Adamson. 1983a. Epithelial layer formation in differentiating aggregates of F9 embryonal carcinoma cells. *J. Cell Biol.* **96**: 1690.

Grover, A., G. Andrews, and E.D. Adamson. 1983b. Role of laminin in epithelium formation by F9 aggregates. *J. Cell Biol.* **97**: 137.

Hogan, B.L.M., A. Taylor, E. Adamson. 1981. Cell interactions modulate embryonal carcinoma cell differentiation into parietal or visceral endoderm. *Nature* **291**: 235.

Ingram, R.S., R.W. Scott, and S.M. Tilghman. 1981. α-Fetoprotein and albumin genes are in tandem in the mouse genome. *Proc. Natl. Acad. Sci. U.S.A.* **78**: 4694.

Kioussis, D., F. Eiferman, P. Van de Rijn, M.B. Gorin, R.S. Ingram, and S.M. Tilghman. 1981. The evolution of α-fetoprotein and albumin II. The structures of the α-fetoprotein and albumin genes in the mouse. *J. Biol. Chem.* **256**: 1960.

Kleinsmith, L.J. and G.B. Pierce, Jr. 1964. Multipotentiality of single embryonal carcinoma cells. *Cancer Res.* **24**: 1544.

Krumlauf, R., R.E. Hammer, S.M. Tilghman, and R.L. Brinster. 1985. Developmental regulation of α-fetoprotein genes in transgenic mice. *Mol. Cell. Biol.* (in press).

Liesi, P., L. Rechardt, and J. Wartiovaara. 1983. Nerve growth factor induces adrenergic neuronal differentiation in F9 teratocarcinoma cells. *Nature* **306**: 265.

Martin, G.R. 1975. Teratocarcinomas as a model system for the study of embryogenesis and neoplasia. *Cell* **5**: 229.

Nadijcka, M. and N. Hillman. 1974. Ultrastructural studies of the mouse blastocyst substages. *J. Embryol. Exp. Morphol.* **32**: 675.

Scott, R.W., T.F. Vogt, M. Croke, and S.M. Tilghman. 1984. Tissue-specific activation of a cloned α-fetoprotein gene during differentiation of a transfected embryonal carcinoma cell line. *Nature* **310**: 562.

Solter, D. and I. Damjanov. 1973. Explantation of extraembryonic parts of 7-day-old mouse egg cylinders. *Experientia* **29**: 701.

Strickland, S. and V. Mahdavi. 1978. The induction of differentiation in teratocarcinoma stem cells by retinoic acid. *Cell* **15**: 393.

Strickland, S., K.K. Smith, and K.R. Marotti. 1980. Hormonal induction of differentiation in teratocarcinoma stem cells: Generation of parietal endoderm by retinoic acid and dibutyryl cAMP. *Cell* **21**: 347.

Tilghman, S.M. and A. Belayew. 1982. Transcriptional control of the murine albumin/α-fetoprotein locus during development. *Proc. Natl. Acad. Sci. USA* **79**: 5254.

Tilghman, S.M., D. Kioussis, M.B. Gorin, J.P. Garcia Ruiz, and R.S. Ingram. 1979. The presence of intervening sequences in the α-fetoprotein gene of the mouse. *J. Biol. Chem.* **254**: 7393.

Nuclear Transfer in Mammalian Embryos: Genomic Requirements for Successful Development

JAMES McGRATH AND DAVOR SOLTER
The Wistar Institute
Philadelphia, Pennsylvania 19104

OVERVIEW

The nuclear transfer technique was used to explore the functional status of the genome that is compatible with normal mammalian development. By transferring a single pronucleus into zygotes from which only one pronucleus was removed, we constructed diploid heterozygous gynogenones (two female pronuclei) and androgenones (two male pronuclei). Such embryos complete preimplantation development, but die before midgestation. These data indicate that the genomes of male and female gametes are functionally different and that both are required for normal development. We also studied the restriction of totipotency of blastomere nuclei by transferring nuclei from successive stages of development into enucleated zygotes. Only nuclei from two-cell stage embryos are able to support preimplantation development although less well than the zygote pronuclei; nuclei from older embryos (four-, eight-cell stage embryos, and inner cell mass) do not support the preimplantation development. The restriction of developmental totipotency occurs very early during mammalian development and probably reflects the need for early and precise activation of embryonic genome.

INTRODUCTION

At the beginning of development, the zygote consists of several different components. The haploid genomes contributed by the egg and the sperm are presumably identical in a quantitative sense with the exception of sex chromosomes. The majority of the extrachromosomal material, i.e., cytoplasmic macromolecules, organelles, and the plasma membrane are derived from the egg; however, some extragenomic contribution from the sperm likely exists. The ability to experimentally recombine and manipulate different components of the zygote and to observe the effect of such manipulations on development should provide a powerful approach to the study of mechanisms controlling development. Probably the most studied of the various possible manipulations is development without the genetic, and usually without the extragenetic contribution of the male gamete, i.e., parthenogenesis. Successful parthenogenetic development occurs in nature within most animal phyla and in all classes of vertebrates except mammals (Beatty 1967; Cuellar 1977; Markert 1982). It is likely that this type of development, though evolutionarily less adaptable, provides an advantage in a particular ecological situation (Cuellar 1977). Considering the prevalence of parthenogenesis among animal species, it is puzzling why, despite numerous experimental attempts, no confirmed parthenogenones were ever observed in mammals. Most mammalian parthenogenones die soon after implantation (Graham 1974) and a very few persist until midgestation (Kaufman et al. 1977). Both nuclear (homozygosity for lethal genes) and cytoplasmic (lack of extragenomic contribution of the sperm, or the absence of a proper fertilization stimulus) deficiencies have been proposed to explain the developmental failure of parthenogenetic embryos. To provide a framework for understanding the lethality of parthenogenetic and uniparental embryos, we have transferred single pronuclei between zygotes determining if the maternal and paternal genomic contributions are equivalent (McGrath and Solter 1984a).

In parallel investigations we have queried if somatic cell nuclei can replace the zygote nucleus to determine if they regain their totipotency within the zygote cytoplasm. This approach should provide information as to whether development and differentiation leads to irreversible quantitative and qualitative changes of that portion of the genome necessary to complete development. The most detailed analyses of this question were conducted using amphibians and, although some questions remain, a few basic conclusions were made. Embryonic amphibian nuclei transferred to zygotes whose nuclei were destroyed can support complete development (Briggs and King 1952; Gurdon 1962a), although this ability gradually diminishes as development progresses (Gurdon 1962b); thus, nuclei transplanted from adult cells are able to support only minimal embryonic development (Gurdon 1974), but never development to the adult. Staging the restriction of developmental potency of the nucleus provides an indication of when the changes in the genome, which regulate or are concomitant with differentiation, occur. Recently these sort of experiments have been attempted in mammals (Illmensee and Hoppe 1981). Detailed analysis of the developmental potential of nuclei from successive stages of preimplantation development have been completed (McGrath and Solter 1984b) and some of these results are presented here.

The experimental procedures used in the following experiments consist of removal of one or both pronuclei from the zygote and the transfer of different nuclei into the enucleated egg (McGrath and Solter 1983a,b, 1984a). Intact nuclei are removed within a cytoplasmic vesicle, without penetration of the plasma membrane of the egg or blastomere, using a large bore pipette. Such karyoplasts are then placed under the zona pellucida of previously enucleated zygotes and the karyoplasts and cytoplasts are fused using inactivated Sendai virus. This procedure ensures complete removal of the nuclear material, and is essentially nontraumatic, thus providing greater than 90% survival of manipulated eggs (McGrath and Solter 1983a).

RESULTS

Development of Embryos Containing Only the Maternal or Only Paternal Genome

Paternal and maternal pronuclei can be distinguished in the zygote on the basis of their position. Replacement of one pronucleus with the pronucleus of the opposite sex results in embryos with two female pronuclei (biparental gynogenones) or two male pronuclei (biparental androgenones). Such embryos are immediately placed into the oviducts of pseudopregnant foster mothers and the resulting progeny examined. By the use of two donor strains with distinguishable genetic markers, it is always possible to identify the isoparental individuals from those in which the maternal/parental composition of the genone is maintained. Control embryos in these experiments were produced by replacing one pronucleus with the pronucleus of the same parental sex from another zygote. These control embryos develop to term (Table 1). When the development of a large series of androgenetic and gynogenetic embryos was analyzed (Table 1), only animals that were derived from embryos with a maternal/paternal nuclear composition were born. The appearance of animals of the maternal/paternal phenotype that are born following transfer of presumed gynogenetic and androgenetic embryos indicate that the visual assignment of the origin of pronuclei in the zygote was incorrect in some cases. On the basis of the number of control progeny born, we would expect to have found 19 gynogenones and 14 androgenones; no progeny of such phenotype were observed ($X^2 = 27.15; p < 0.001$). Therefore, these results indicate that diploid embryos containing two maternal or two paternal genomes do not complete development. Since all experimental embryos were derived from fertilized zygotes and nearly half of them were heterozygous (because two different inbred strains were used), we conclude that neither absence of the extragenomic sperm contribution nor excessive

Table 1
Progeny Resulting from Biparental Gynogenetic and Androgenetic and Control Zygotes[a]

Presumed genotype of the zygote	No. of embryos transferred	No. of progeny	Genotype consistent with the observed phenotype of the progeny
Maternal/maternal (gynogenones)	339	3	Maternal/paternal
Paternal/paternal (androgenones)	328	2	Maternal/paternal
Maternal/paternal	348	18	Maternal/paternal

[a]Data presented are summarized from McGrath and Solter (1984a). Embryos from various crosses were used so that the resulting gynogenones and androgenones were homozygous in some experiments and heterozygous in others. The control embryos (in which the removed pronucleus was replaced with the pronucleus of the same parental sex) were genotypically the reciprocal of experimental embryos.

homozygosity can explain the death of these embryos. Our previous investigation has shown that preimplantation development and the rate of implantation of androgenones and gynogenones is comparable to that of control embryos; however, following implantation these embryos soon display signs of growth retardation and degeneration and all die by midgestation (McGrath and Solter 1984a).

Development of Enucleated Zygotes to Which Blastomere Nuclei Are Transferred

To assess the ability of preimplantation stage nuclei to support development, karyoplasts from zygote, two-, four-, and eight-cell stage embryos and single inner cell mass (ICM) cells were introduced into enucleated zygotes as previously described (McGrath and Solter 1983b, 1984b). The resultant embryos were subsequently cultured for 5 days in vitro. (Tables 2–4 include data previously reported by McGrath and Solter [1984b] plus approximately 100 additional nuclear transplant embryos.) The results (Table 2) show that whereas a high proportion of enucleated zygotes receiving pronuclei from a second zygote successfully complete preimplantation development, relatively few of the embryos receiving later-stage nuclei do so. Although these latter embryos frequently divide to the two-cell stage, subsequent development is significantly impaired. Approximately 15% of embryos receiving two-cell stage nuclei developed to the morula-blastocyst stages. A small proportion of embryos (5%) receiving four-cell stage nuclei developed to the early morula stage; blastocelic cavity formation by these embryos, however, was impaired and resulted in embryos with multiple, small, fluid-filled cavities. Embryos receiving eight-cell stage or ICM cell nuclei only infrequently developed beyond the two-cell stage and never achieved the morula-blastocyst stages. Additionally, embryos receiving ICM cell nuclei often exhibited marked distortions of the embryo plasma membrane and unequal segregation of a majority of cytoplasm to one of the two blastomeres.

The inability of nuclear transplant embryos receiving late preimplantation stage nuclei to develop may result from the inability of the transferred nucleus to function appropriately or from the simultaneous transfer of foreign cytoplasm. Karyoplasts from two-, four-, and eight-cell stage embryos and ICM cells were therefore introduced into nonenucleated zygotes. The data (Table 3) show that a greater proportion of nonenucleated zygotes receiving later-stage nuclei successfully develop in vitro than do enucleated embryos. This result is consistent with a previous report which demonstrated complete preimplantation development of nonenucleated zygotes receiving ICM cell nuclei (Modlinski 1981). Thus, the presence of a foreign nucleus or cytoplasm alone will not completely block development (although one or both may be inhibitory). To assess a possible deleterious effect resulting from the transfer of foreign cytoplasm, a small volume of membrane bound cytoplasm was removed from zygotes and replaced with membrane bound cytoplasm from two-, four-, and eight-cell stage embryos. The results (Table 4) show that the intro-

Table 2
In Vitro Development of Enucleated One-cell Stage Embryos Receiving Cleavage Stage Nuclei

Nuclear donor	Total no. embryos	Development					
		Zygote	Two-cell	Three-cell	Four-eight-cell	Morula	Blastocyst
Zygote	25	0 (0)[a]	1 (4)	0 (0)	0 (0)	0 (0)	24 (96)
Two-cell	191	59 (31)	66 (35)	21 (11)	15 (8)	10 (5)	20 (10)
Four-cell	87	21 (24)	59 (68)	1 (1)	2 (2)	4 (5)	0 (0)
Eight-cell	120	34 (28)	85 (71)	0 (0)	(1)	0 (0)	0 (0)
ICM cell	84	38 (45)	45 (54)	0 (0)	1 (1)	0 (0)	0 (0)
Control	283	8 (3)	8 (3)	2 (1)	14 (5)	32 (11)	219 (77)

[a]Number of embryos not developing beyond given developmental stage; ( ) refers to percentage of embryos arresting at given developmental stage/total number of embryos.

Table 3
In Vitro Development of Nonenucleated One-cell Stage Embryos Receiving Cleavage Stage Nuclei

Nuclear donor	Total no. embryos	Development				
		Zygote	Two-three-cell	Four-eight-cell	Morula	Blastocyst
Two-cell	45	5 (11)[a]	2 (4)	1 (2)	7 (16)	30 (67)
Four-cell	15	2 (13)	4 (27)	1 (7)	1 (7)	7 (47)
Eight-cell	38	4 (11)	11 (29)	8 (21)	8 (21)	9 (24)
ICM cell	20	2 (10)	7 (35)	2 (10)	4 (20)	5 (25)
Control	61	0 (0)	3 (5)	1 (2)	5 (8)	52 (85)

[a]Number of embryos not developing beyond given developmental stage; ( ) refers to percentage of embryos arresting at given developmental stage/total number of embryos.

duction of foreign cytoplasm into nonenucleated zygotes did not result in a significant decrease in development.

The results show that nuclei obtained from successive mouse embryo cleavage stages rapidly lose their ability to support preimplantation development when introduced into enucleated zygotes. We infer that late preimplantation stage nuclei are unable to sufficiently reprogram their nuclear activity to accommodate the transcriptional requirements of the early cleavage stage embryo. Inhibition of RNA synthesis results in developmental arrest of mouse embryos at the two-cell stage, thus, if late preimplantation stage nuclei (i.e., eight-cell or ICM cell) are unable to produce two-cell stage-specific transcripts, development of these nuclear transplant embryos would be halted. Data consistent with this suggestion have been obtained by removing the nuclei of both blastomere from late two-cell stage embryos and fusing this cytoplast with an eight-cell stage karyoplast. Although possessing a reduced cell number, these embryos develop to the blastocyst stage (J. McGrath and D. Solter, unpubl.). Thus, whereas enucleated zygotes receiving eight-cell stage nuclei do not complete preimplantation development, complete enucleated two-cell embryos receiving similar stage nuclei do. It remains to be determined, however, whether enucleated two-cell stage embryos receiving even later stage embryonic nuclei, e.g., ICM cell, also develop.

Table 4
In Vitro Development of Nonenucleated One-cell Stage Embryos Receiving Cytoplasm from Cleavage Stage Embryos

Cytoplasmic donor	Total no. embryos	Development				
		Zygote	Two-three cell	Four-eight cell	Morula	Blastocyst
Two-cell	8	0 (0)[a]	0 (0)	2 (25)	1 (13)	5 (63)
Four-cell	14	0 (0)	0 (0)	0 (0)	0 (0)	14 (100)
Eight-cell	25	3 (12)	2 (8)	0 (0)	3 (12)	17 (68)

[a]Number of embryos not developing beyond given developmental stage; ( ) refers to percentage of embryos arresting at given developmental stage/total number of embryos.

To assess the response of differentiated nuclei to zygote cytoplasm, ICM and teratocarcinoma cells (line PCC3) were fused with enucleated zygotes and the resulting embryos incubated in vitro for various lengths of time. The embryos (one- and two-cell stages) were fixed, stained with aceto-lacmoid (Toyoda and Chang 1974) and examined using light microscopy. Considerable nuclear swelling occurs after 6 hours in culture (Fig. 1). Metaphase plates can be observed in some one-cell stage embryos after 24 hours in vitro (Fig. 2C). Two-cell stage embryos derived from fusion with ICM or teratocarcinoma cells have nuclei in both blastomeres (Fig. 2B, D), however, the large nucleoli observed in control two-cell stage embryo (Fig. 2A) are either very much reduced in number (Fig. 2B) or completely absent (Fig. 2D). The significance of this observation for the developmental failure of nuclear transfer embryos is unclear. These data indicate that the transferred differentiated nuclei do respond to the egg cytoplasm by swelling and are also able to participate in apparently normal division.

The results of the present studies contrast with reports of complete development of nuclear transplant mouse embryos receiving ICM cell nuclei (Illmensee and Hoppe 1981; Hoppe and Illmensee 1982). Although the experiments described above transferred nuclei via karyoplast fusion rather than mechanical injection, the introduction of ICM cell nuclei into enucleated zygotes via mechanical transfer in the authors' laboratory did not result in improved development (McGrath and Solter 1984b). A possible cause for this apparent discrepancy has been detailed elsewhere (McGrath and Solter 1984a,b).

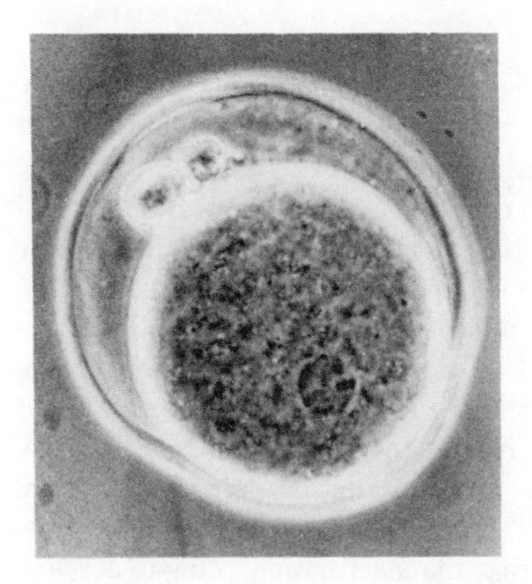

Figure 1
Enucleated mouse zygote fused with PCC3 embryonal carcinoma cell and fixed 6 hrs after fusion. Notice the swelling of the transferred nucleus. (×450).

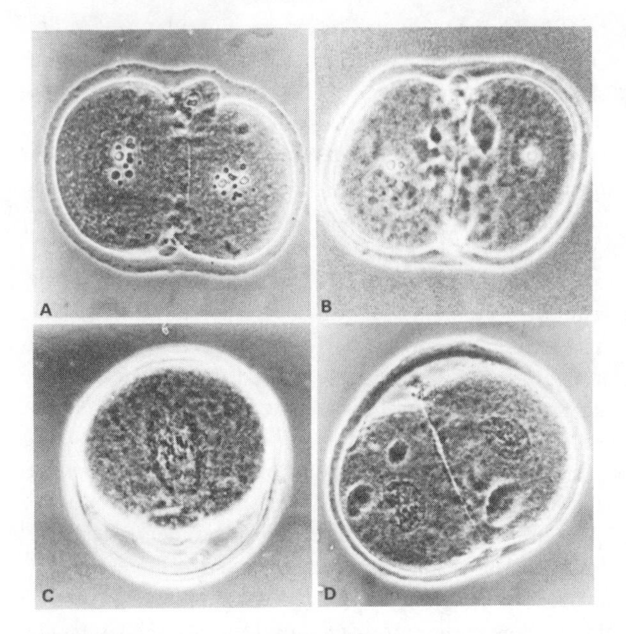

Figure 2
(*A*) Zygote isolated and grown in vitro for 24 hrs. (*B*) Enucleated zygote fused with ICM cell and grown in vitro for 24 hrs. (*C*) and (*D*) Enucleated zygotes fused with PCC3 embryonal carcinoma cells and grown for 24 hrs. Notice the metaphase plate (*C*) and several cytoplasmic inclusions (*B, D*). Large and numerous nucleoli are visible in control two-cell stage embryos (*A*). Very few such nucleoli are seen following transfer of ICM cell nucleus (*B*) and none following transfer of teratocarcinoma cell nucleus (*D*). (×450).

DISCUSSION

It is apparent from the results presented that the process of mammalian development is very sensitive to changes of the basic components of the fertilized egg. It seems that qualitative changes have more drastic consequences than the quantitative changes. Quantitative changes of the genome result in triploid, quadruploid, and trisomic embryos, which undergo substantial development and some trisomic embryos can develop to term (Magnuson and Epstein 1981; Gropp et al. 1983). Development can proceed normally, if, for example, part of the egg cytoplasm is removed or if the size of the embryo is doubled by aggregation of two preimplantation embryos. In contrast, qualitative changes, like replacement of the pronucleus of one parent with the pronucleus of the parent of the opposite sex, or mismatching the developmental stage of the cytoplasm and the nucleus, are embryo lethal despite the fact that these changes do not affect the normal amount of genetic material or of the cytoplasm. It is very likely that the mechanism leading to developmental failure of diploid androgenones and gynogenones are different from those which operate in zygotes containing nuclei from cleavage stage embryos.

Our results (McGrath and Solter 1984a) indicate that the paternal and maternal genome are not identical and that neither of them alone is sufficient for development. Very similar data concerning developmental failure of gynogenones were also obtained by Surani et al. (1984), and it is likely that spontaneous and experimentally induced parthenogenetic embryos also fail because of absence of the male genome. The difference between the male and female genome at the time of fertilization is probably functional and both genomes have to become identical at least in primordial germ cells, so that during the first meiotic division, completely random segregation of maternal and paternal chromosomes can occur. It is probable that the difference between the male and female genome is re-established during gametogenesis. There are several examples of the male and female genomes functioning differently during development. The paternal X chromosome is preferentially inactivated in extraembryonic membrane (Takagi and Sasaki 1975) and the proximal part of chromosome 17 behaves differently depending upon whether it is derived from the mother or from the father (Johnson 1975; Lyon and Glenister 1977; McGrath and Solter 1984c). It is very likely that such differences will be observed for other parts of the genome as well.

We suggested (McGrath and Solter 1984a) that the need for differential functioning of maternal and paternal genome during development logically precludes the possibility of cloning, since it would be very difficult to re-establish exactly the same type of differential activity using somatic cell nuclei. This might be the case, but some other factors must also play a role since embryos containing blastomere nuclei die much sooner than androgenetic and gynogenetic embryos. Actually, except for embryos produced by transfer of two-cell stage nuclei, the embryos containing blastomere nuclei arrest after one or a few divisions. Such early failure is probably related to the basic characteristics of the mammalian embryo. Cleavage divisions and early development in nonmammalian species is independent of the activation of the embryonic genome (Davidson 1976). On the other hand, the embryonic genome in the mouse is expressed very early, possibly in the zygote and certainly in the two-cell stage embryo. Synthesis of polyadenylated RNA is detected in zygotes (Clegg and Piko 1982, 1983), the synthesis of some new proteins is observed in the zygote (Bensaude et al. 1983), but the major change in the polypeptide profile occurs at the two-cell stage (Levinson et al. 1978; Howe and Solter 1979; Flach et al. 1982; Bolton et al. 1984). It is easy to imagine that some of these newly synthesized polypeptides are stage-specific and essential for normal development and that corresponding genes from older nuclei are not reactivated following transfer. If this is indeed the case, one might achieve successful development by transferring blastomere nuclei into slightly older (two- or four-cell stage) cytoplasm; our preliminary data supports such a notion. It is impossible to predict whether, after successfully removing the first obstacle to cloning and achieving preimplantation development, we will encounter another obstacle and the embryos will fail because the differential functioning of maternal and paternal genome was not established from the beginning.

It is also worth emphasizing that the nucleus-cytoplasm mismatch in the opposite direction, namely transfer of developmentally younger nuclei into older cytoplasm, may also result in developmental failure. We have so far tried only transfer of pronuclei into the two-cell stage cytoplasm and these embryos failed to develop. This result implies that something in the cytoplasm of the older embryos adversely affects the transferred nucleus, or that the asynchrony of genomic activity and the stage of the cytoplasm is deleterious for the embryo. It is obvious that very precise nucleo-cytoplasmic interactions are required for successful mammalian development.

The experiments presented here point to the relative uniqueness of mammalian development. Early and correct activation of the embryonic genome is mandatory, and nuclear transfer experiments will provide information as to the precise timing of such activation, thus enabling us to search for and identify genes that regulate development.

ACKNOWLEDGMENTS

This work was supported by Grants HD-12487 and HD-17720 from the National Institute of Child Health and Human Development and by Grants CA-10815 and CA-25875 from the National Cancer Institute. The thoughtful comments of Dr. Barbara B. Knowles are gratefully acknowledged.

REFERENCES

Beatty, R.A. 1967. Parthenogenesis in vertebrates. In *Fertilization* (ed. C.B. Metz and A. Monroy), vol. 1, p. 413. Academic Press, New York.

Bensaude, O., C. Babinet, M. Morange, and F. Jacob. 1983. Heat shock proteins, first major products of zygotic gene activity in mouse embryo. *Nature* **305**: 331.

Bolton, V.N., P.G. Oades, and M.H. Johnson. 1984. The relationship between cleavage, DNA replication, and gene expression in the mouse 2-cell embryo. *J. Embryol. Exp. Morphol.* **79**: 139.

Briggs, R. and T.J. King. 1952. Transplantation of living nuclei from blastula cells into enucleated frogs' eggs. *Proc. Natl. Acad. Sci.* **38**: 455.

Clegg, K.B. and L. Piko. 1982. RNA synthesis and cytoplasmic polyadenylation in the one-cell mouse embryo. *Nature* **295**: 342.

——— . 1983. Quantitative aspects of RNA synthesis and polyadenylation in 1-cell and 2-cell mouse embryos. *J. Embryol. Exp. Morphol.* **74**: 169.

Cuellar, O. 1977. Animal parthenogenesis. *Science* **197**: 837.

Davidson, E.M. 1976. *Gene activity in early development,* 2nd ed. Academic Press, New York.

Flach, G., M.H. Johnson, P.R. Braude, R.A.S. Taylor, and V.N. Bolton. 1982. The transition from maternal to embryonic control in the 2-cell mouse embryo. *Eur. Mol. Biol. Organ. J.* **1**: 681.

Graham, C.F. 1974. The production of parthenogenetic mammalian embryos and their use in biological research. *Biol. Rev. Camb. Philos. Soc.* **49**: 399.

Gropp, A., H. Winking, E. W. Herbst, and C.P. Claussen. 1983. Murine trisomy: Developmental profiles of the embryo, and isolation of trisomic cellular systems. *J. Exp. Zool.* **228**: 253.

Gurdon, J.B. 1962a. Adult frogs derived from the nuclei of single somatic cells. *Dev. Biol.* **4**: 256.

——— . 1962b. The developmental capacity of nuclei taken from intestinal epithelium cells of feeding tadpoles. *J. Embryol. Exp. Morphol.* **10**: 622.

——— . 1974. *The control of gene expression in animal development.* Clarendon Press, Oxford.

Hoppe, P.C. and K. Illmensee. 1982. Full-term development after transplantation of parthenogenetic embryonic nuclei into fertilized mouse eggs. *Proc. Natl. Acad. Sci. U.S.A.* **79**: 1912.

Howe, C.C. and D. Solter. 1979. Cytoplasmic and nuclear protein synthesis in preimplantation mouse embryos. *Dev. Biol.* **52**: 209.

Illmensee, K. and P.C. Hoppe. 1981. Nuclear transplantation in Mus musculus: Developmental potential of nuclei from preimplantation embryos. *Cell* **23**: 9.

Johnson, D.R. 1975. Further observations on the hairpintail (T^{hp}) mutation in the mouse. *Genet. Res.* **24**: 207.

Kaufman, M.H., S.C. Barton, and M.A.H. Surani. 1977. Normal postimplantation development of mouse parthenogenetic embryos to the forlimb bud stage. *Nature* **265**: 53.

Levinson, J., P. Goodfellow, M. Vadeboncoeur, and H. McDevitt. 1978. Identification of stage-specific polypeptides synthesized during murine preimplantation development. *Proc. Natl. Acad. Sci.* **75**: 3332.

Lyon, M.F. and P.H. Glenister. 1977. Factors affecting the observed number of young resulting from adjacent-2 disjunction in mice carrying a translocation. *Genet. Res.* **29**: 83.

Magnuson, T. and C.J. Epstein. 1981. Genetic control of very early mammalian development. *Biol. Rev. Camb. Philos. Soc.* **56**: 369.

Markert, C.L. 1982. Parthenogenesis, homozygosity, and cloning in mammals. *J. Hered.* **73**: 390.

McGrath, J. and D. Solter. 1983a. Nuclear transplantation in the mouse embryo by microsurgery and cell fusion. *Science* **220**: 1300.

——— . 1983b. Nuclear transplantation in mouse embryos. *J. Exp. Zool.* **228**: 355.

——— . 1984a. Completion of mouse embryogenesis requires both the maternal and paternal genomes. *Cell* **37**: 179.

——— . 1984b. Inability of mouse blastomere nuclei transferred to enucleated zygotes to support development in vitro. *Science* **226**: 1317.

——— . 1984c. Maternal T^{hp} lethality in the mouse is a nuclear, not cytoplasmic, defect. *Nature* **308**: 550.

Modlinski, J.A. 1981. The fate of inner cell mass and trophectoderm nuclei transplanted to fertilized mouse eggs. *Nature* **292**: 342.

Surani, M.A.H., S.C. Barton, and M.L. Norris. 1984. Development of reconstituted mouse eggs suggests imprinting of the genome during gametogenesis. *Nature* **308**: 548.

Takagi, N. and M. Sasaki. 1975. Preferential inactivation of the paternally derived X-chromosome in the extraembryonic membranes of the mouse. *Nature* **256**: 640.

Toyoda, Y. and M.C. Chang. 1974. Fertilization of rat eggs in vitro by epididymal spermatozoa and the development of eggs following transfer. *J. Reprod. Fertil.* **36**: 9.

Regulation of Embryogenesis by Maternal and Paternal Genomes in the Mouse

AZIM SURANI
AFRC Institute of Animal Physiology
307 Huntingdon Road
Cambridge CB3 0JQ, United Kingdom

OVERVIEW

The role of maternal and paternal genomes during development, when examined by manipulating the genetic constitution of eggs and blastocysts, demonstrates that both are needed for development to term. However, the maternal genome is sufficient for development to the blastocyst stage and to some extent up to the 25-somite midgestation stage. By contrast, the paternal genome is essential for the proliferation of the trophoblast although it is less efficient in directing preimplantation development. Hence substantial improvement in the development of the inner cell mass from parthenogenetic embryo occurs when it is introduced into normal trophectoderm. Our results indicate differential but not exclusive roles for the maternal and paternal genomes in the development of the embryo and extraembryonic tissues, respectively. The influence is probably achieved as a consequence of differential imprinting of the genome during gametogenesis. There are indications to suggest that the information is not used for a specific event during embryogenesis but rather throughout development. It is likely that the imprinting is reversed in the primordial germ cells prior to the reinitiation of the process during gametogenesis.

INTRODUCTION

Recent investigations suggest that gametes may differ in more ways than their striking morphological differences. Mature gametes are transcriptionally inactive. The oocytes are large and stockpiled with macromolecules for use immediately after fertilization, and they are arrested at metaphase of second meiotic division. By contrast, spermatozoa are formed after completion of meiosis and they are designed to introduce highly condensed paternal DNA bound with basic protamine-like proteins into oocytes (Wolgemuth 1983).

After fertilization the maternal and paternal genomes are brought together inside the egg cytoplasm and the three components interact during the course of embryogenesis to produce viable young. The egg first completes meiosis by extruding the second polar body containing a haploid set of chromosomes. Male and female pronuclei are then formed and at this time histones are added to sperm DNA after the release of protamines (Wolgemuth 1983). The very early events of development before the embryonic genome becomes activated are probably dictated by stored maternal products. It is generally believed that this occurs as early as the two-cell stage in the mouse (Van Blerkom and Brockway 1975; Flach et al. 1982; Bensaude et al. 1983) although there is even some evidence for transcription at the pronuclear stage (Clegg and Piko 1982). The products from both maternal and paternal genomes have first been detected at the two-cell stage as judged by the synthesis of two genetic variants of β_2-microglobulins (Sawicki et al. 1981). The cytoplasm of early eggs and embryos is therefore continuously modified and the synthesis of new components comes increasingly under the direction of the embryonic genome. Changes in the cytoplasmic properties have also been shown to influence transcriptional activity of donor nuclei in early blastomere (Bernstein and Mukherjee 1972).

Whatever the chronology of early events following fertilization, it has generally been assumed that once the embryonic genome becomes active, the origin of the various chromosomes is irrelevant to their functions and they are genetically equivalent. However, recent investigations suggest that this may not be so (McGrath and Solter 1984a; Surani et al. 1984). The gametes appear to carry information as to the paternal or maternal origin of the genes they express. Such information is likely to be reversibly imprinted into each parental genome during oogenesis and spermatogenesis. Thus it appears that the maternal and paternal genomes may have differential roles during development.

An initial approach to the understanding of the roles of maternal and paternal genomes is to reconstitute eggs to contain only maternal or paternal genomes; the subsequent development of such manipulated eggs can then be assessed. Further insight into the differential role of the two genomes can be gained by reconstruction of blastocysts using inner cell mass and trophectoderm from embryos of different genetic constitution.

RESULTS

Development of embryos lacking a male genome as well as any extragenetic contribution from spermatozoa can be determined by artificial activation of eggs using one of a wide variety of stimuli which initiate the program of development in vivo and in vitro; and activation can also occur spontaneously in the LT/Sv strain of mice (Stevens 1975). Activation is accompanied by polar body extrusion, giving a haploid parthenogenone, or polar body retention in the presence of cytochalasin B to give a diploid parthenogenone (Niemierko 1975; Borsuk 1982). However, although development is initiated, it does not proceed to term (Graham 1974; Stevens 1975; Tarkowski 1975; Kaufman 1978; Whittingham 1980). It has often been argued that this failure is due either to the homozygosity of parthenogenones, which permits the expression of recessive lethal genes, or that it arises from the defective cytoplasm of activated eggs, lacking as it does, an extragenetic contribution from spermatozoa (Graham 1974). Nevertheless, parthenogenetic embryos can develop up to midgestation yielding embryos at the forelimb bud stage with about 25 somites (Kaufman et al. 1977). Moreover, parthenogenetic embryos will develop to term as chimeras when combined with normal embryos (Stevens et al. 1977; Surani et al. 1977) and sometimes will give rise to viable germ cells (Stevens 1978). However, such chimeras do not develop to term if both embryos from different strains of mice are parthenogenetic (M.A.H. Surani and S.C. Barton, unpubl.). These results show that although parthenogenetic embryos develop poorly, they do not suffer from cell lethality.

The results from two studies on activated eggs have led to claims that their cytoplasm was defective and thereby prevented their development to term. First, it was reported that if either the male or the female pronucleus was removed from fertilized eggs and the eggs were then made both diploid and homozygous by suppressing the first cleavage division, a few of the resulting uniparental embryos developed to term (Hoppe and Illmensee 1977). Second, these investigators also claimed that when nuclei from inner cell mass cells of fertilized or parthenogenetic LT/Sv blastocysts were transferred to enucleated fertilized eggs, development to term occurred in a few cases (Hoppe and Illmensee 1982). However, work at several other laboratories has failed to confirm some of these studies (Modlinski 1980; Markert 1982; Surani and Barton 1983). We examined this problem by making use of triploid eggs with two female and one male pronucleus; these eggs were created by suppressing second polar body extrusion in fertilized eggs with cytochalasin B (Niemierko 1975; Borsuk 1982; Surani and Barton 1983). In these experiments the normal genetic constitution can be restored by removing a female pronucleus from the triploid eggs, which thus serves as an internal control for any failure of development attributable to technical causes such as micromanipulation. In contrast, removal of the male pronucleus results in eggs with two female pronuclei (digynic or gynogenetic). Such embryos are similar in their genetic constitution to diploid parthenogenones, differing only in that, being prepared following fertilization, they presumably contain normal cytoplasm. In both outbred and inbred strains of mice (Surani and Barton 1983), including the LT/Sv strain (S.C. Barton and M.A.H. Surani, unpubl.), it was found that development to term occurred only in those triploid eggs from which a female pronucleus was removed to restore a diploid genetic constitution; digynic eggs with two female pronuclei developed poorly and resembled diploid parthenogenetic embryos (Table 1). Hence the failure in this instance cannot be attributable to anomalous cytoplasm as was suggested in the studies cited earlier (Hoppe and Illmensee, 1977, 1982). Furthermore, it was confirmed that development was also possible when pronuclei from fertilized eggs were transferred to enucleated, parthenogenetically activated eggs but not when pronuclei from parthenogenetic eggs were transferred to enucleated fertilized eggs (Mann and Lovell-Badge 1984). The combined results from these experiments indicate that the cytoplasm of activated and fertilized eggs are similar. It is the genotype of digynic and parthenogenetic eggs which appears to be responsible for their nonviability.

Table 1
Manipulation of Triploid Eggs

Strain of mice	Group	Number of eggs		
		Transferred	Implanted	With embryos
MF1	Triploid	78	44	30 Retarded
	Restored diploid from triploids	44	24	16 Normal
	Digynic	77	34	7 Retarded
	Homozygous gynogenetic	95	15	1 Retarded
C57BL/6	Digynic	16	7	1 Retarded
	Homozygous gynogenetic	92	6	0
BALB/C	Digynic	8	7	1 Retarded
	Homozygous gynogenetic	57	3	0
$(C57BL/6XCBA)F_1$	Digynic	36	14	10 Retarded
	Homozygous gynogenetic	20	0	0

Other studies have been carried out to establish whether homozygosity or the origin of pronuclei was the cause of developmental failure in digynic and parthenogenetic embryos (Surani et al. 1984). In one type of experiment parthenogenetic haploid eggs were used as recipient eggs into which a second pronucleus, either male or female, from fertilized eggs of another strain of mice was introduced by Sendai virus-assisted fusion of the karyoplast containing the donor pronucleus (Graham 1971), using the technique of McGrath and Solter (1983). Hence, although the reconstituted eggs were similar genetically, the origin of the donor pronucleus was either paternal or maternal. When the donor pronucleus was male, development proceeded to term. When the donor pronucleus was female, these reconstituted heterozygous eggs with two female pronuclei failed to develop to term and the maximum development attained was about the same as we observed in the case of parthenogenones and digynic embryos (Table 2).

Further studies were carried out to determine how biparental eggs with two male pronuclei (androgenetic) developed (Barton et al. 1984). Heterozygous diploid androgenetic eggs were made by removing the female pronucleus from a fertilized egg and transferring to the egg a second male pronucleus taken from another fertilized egg. Only about 20% of such eggs implanted and even fewer yielded any postimplantation embryonic derivatives. However, about 25% of androgenetic eggs are lost because the eggs with two Y chromosomes cannot cleave more than two or three times (Morris 1968), and it is not known whether there is any difference in the development of androgenetic XX (25%) and XY (50%) embryos. The control eggs for this experiment in which they were reconstituted with a male and a female pronucleus developed to term, in some instances almost 50% of such manipulated eggs producing live young (Table 3).

Taken together these studies argue against differences between the cytoplasms of activated and fertilized eggs being important in the failure of parthenogenones and digynic embryos. Moreover, failure can no longer be ascribed entirely to homozygosity since biparental heterozygous eggs with two female or two male pronuclei also failed to develop to term. It appears that a paternal and a maternal genome must both be present in an egg for development to proceed to term. Hence paternal and maternal genomes must carry "information" regarding their origin, and this information is crucial for development.

It is not entirely clear how paternal and maternal genomes differ, but a careful examination of embryos provided some clues about their differential roles during embryogenesis. There are some marked differences between development of gynogenetic and androgenetic eggs. A total of 80% or more of the parthenogenetic and gynogenetic eggs develop to the blastocyst stage whereas only about 20% of androgenetic eggs do so on average; and larger numbers of parthenogenetic and gynogenetic eggs therefore also implant compared to androgenetic eggs. The most interesting differences in development are observed on day 11 of pregnancy (length of gestation in mice is 19 days) after which time the embryos deteriorate. The most advanced parthenogenetic and gynogenetic embryos obtained on day 10-11 of

Table 2
Development of Parthenogenetic Eggs with a Male or a Female Donor Pronucleus

Type of reconstituted egg[a]	No. of eggs[c] transferred	Embryos[d]				
		Implanted	Normal	Small	Retarded abnormal	None
Pn + Ak MF1	24	–	9 live young[b]			
Pn + Gk MF1	48	38	0	8	12	18
Pn + Gk F_1	24	12	0	0	4	8
Pn + Pnk	7	4	0	0	0	4
Unoperated MF1 × MF1	106	88	65	4	2	17

[a]Pn = Haploid parthenogenone: Ak = Male (androgenetic) pronucleus; Gk = Female (gynogenetic) pronucleus; F_1 = (C57BL/6 × CBA)F_1: Non albino: Gpi bb; MF1 = Albino: Gpi aa
[b]Five females and four males
[c]Data from successful recipients only
[d]On day 10 of pregnancy

Table 3
Development of Reconstituted Control and Androgenetic Eggs

Type of reconstituted eggs[a]	Number of eggs		
	Transferred	Implanted	With embryos
Unoperated MF1 × MF1 and $F_1 \times F_1$	190	122	102 Normal
Gk F_1 + Ak F_1	75	–	35 Live young
Ak MF1 + Gk MF1	71	–	27 Live young
Total number of control eggs with a male and a female pronucleus	146		62 Live young
Ak MF1 + Ak F_1	90	20	6
Ak MF1 + Ak MF1	18	5	2
Ak F_1 + Ak MF1	11	1	0
Ak F_1 + Ak F_1	3	2	0
Total number of androgenetic eggs	122	28	8 Retarded[b]

[a]Ak = Androgenetic (male): Gk = gynogenetic (female); F_1 = (C57BL × CBA)F_1 : Non albino: MF1 = Albino, Gpi aa Gpi bb
[b]On day 10–11 of pregnancy

pregnancy have reached the 25-somite stage with vestigial limbs and beating heart, but they all tend to be small (Kaufman et al. 1977; Surani and Barton 1983; Surani et al. 1984). The other characteristic feature of these embryos is the poor development of extraembryonic tissues: the yolk sac is small with particularly meager vasculature and the ectoplacental cone and trophoblast are very sparse (Surani et al. 1984). It can be concluded that development directed entirely by the stored maternal products in the egg and the maternal genome alone can proceed through advanced stages of embryogenesis but appears to fail when rapid growth of the embryo is needed after day 11. Perhaps the poorly developed extraembryonic tissue is partly responsible at a time when the embryos' reliance on the membranes for nutrition increases (Surani et al. 1984). By contrast, the few biparental androgenetic embryos so far obtained display a different morphology on day 11, i.e., these embryos have relatively substantial ectoplacental cone and trophoblast but a very poorly developed embryo (Barton et al. 1984). It is also relevant to note that in the human, androgenetic embryos sometimes develop into hydatidiform moles with extensive proliferation of chorionic vesicles and placenta but the embryo is usually absent (Bagshawe and Lawler 1982). The gross differences in development of reconstituted embryos imply differential or asynchronous gene expression dependent on whether the origin of the genome is maternal or paternal.

Because the apparent difference between the development of the embryo and the extraembryonic tissues depends on the genotype of the embryo, we have started experiments to reconstruct blastocysts as described by Gardner (1972) using the inner cell mass (ICM) and trophectoderm of different genetic constitutions. Because of the ready availability of parthenogenetic and normal blastocysts, the initial experiments have been conducted using the two components of these embryos (M.A.H. Surani and S.C. Barton, unpubl.). In control experiments it was found that when normal ICMs were introduced into normal trophectoderm, over 50% of such reconstituted embryos developed to term. However, when normal ICMs were introduced into parthenogenetic trophectoderm, embryonic development resembled that obtained with unoperated parthenogenetic embryos and a few, small, 25-somite embryos were obtained in which the trophoblast failed to proliferate. By contrast, when parthenogenetic ICMs were introduced into normal trophectoderm, there was a marked improvement of development of parthenogenetic embryos that were larger than the reconstituted embryos described above and all of them developed to at least 25-somite stage embryos, instead of the usual 10-20% (Table 4). Furthermore, we have now obtained more advanced parthenogenetic embryos on day 12 of pregnancy whereas none were obtained previously at this stage of gestation. In such reconstituted blastocysts, the trophoblast proliferates just as in the control embryos, but the yolk sac, which is derived from the inner cell mass, appears to be poorly developed with meager blood supply. Further experiments are in progress to reconstitute blastocysts using gynogenetic and androgenetic embryos. We also plan to introduce normal endoderm cells as well as trophectoderm cells so that perhaps both the yolk sac and the trophoblast proliferation can be restored to normal levels.

Table 4
Development of Reconstituted Blastocysts

Source of[a]				Embryos[c]	
Inner cell mass	Trophectoderm	Number transferred[b]	Implanted	Small with 25 Somites[b]	Very retarded
Fertilized	Fertilized	16	13	7 Normal	
Parthenogenetic	Fertilized	21	19	14	–
Fertilized	Parthenogenetic	39	29	3	10

[a] (C57BL × CBA) F_1 : Gpi bb and CFLP: Gpi aa mice used.
[b] Data from successful recipients only
[c] Between days 10–14 of pregnancy

DISCUSSION

These studies demonstrate that both a paternal and a maternal genome are essential for development to term in the mouse (Barton et al. 1984; McGrath and Solter 1984a; Surani et al. 1984). The maternal genome is sufficient for preimplantation development to the blastocyst stage and to some extent up to the 25-somite midgestation stage (Surani and Barton 1983; Surani et al. 1984). The paternal genome is essential for the proliferation of extraembryonic tissues, especially the trophoblast, although it is less effective in directing preimplantation development (Barton et al. 1984). However, the results do not indicate exclusive roles of the two genomes in the development of the embryo and extraembryonic tissues.

There was a substantial improvement in the development of parthenogenetic embryos when provided with normal trophectoderm in reconstituted blastocysts (M.A.H. Surani and S.C. Barton, unpubl.). These reconstructed embryos have not yet reached term despite extensive proliferation of the trophoblast. However, this may be partly attributed to the poor development of the yolk sac which may need improvement to achieve development to term. Alternatively embryos lacking a paternal genome may already have attained their maximum developmental potential. Further studies on reconstructed blastocysts using inner cell mass and trophectoderm from gynogenetic and androgenetic embryos should provide decisive information. However, from our preliminary studies on aggregation chimeras, it appears that the rescue of nonviable embryos requires the presence of some cells from normal fertilized embryos (M.A.H. Surani and S.C. Barton, unpubl.). A conceptus may not survive if it consists of a mixture of cells without at least some cells from normal fertilized embryos being present.

The apparent differences between maternal and paternal genomes could result from specific imprinting of the genomes during gametogenesis (Surani et al. 1984; Surani 1984). However, it is not known how such information is introduced nor how it is carried. It could take the form of proteins bound to the DNA of gametes or pronuclei, or the DNA itself may be differently modified. The fact that a dispersed, repetitive DNA sequence in sperm DNA is hypermethylated but is undermethylated in oocytes (Sanford et al. 1984) shows that, in principle, these types of differences can occur and that they may influence subsequent gene expression.

Whatever the nature of the information introduced by imprinting of the genomes, it does not appear to be used for a specific event during embryogenesis, but rather it continuously influences various events during development. Studies on mouse mutants, for example, show that the effects on embryogenesis are diverse and determined by the paternal or maternal origin of the inheritance. In the DDK strain of mice, females mated to alien males have low fertility, but not when DDK males mate with alien females (Wakasugi 1974). In the hairpin-tail (T^{hp}) mutation, the heterozygote is viable when the T^{hp} mutation is derived from the father but lethal at about midgestation stage when derived from the mother (Johnson 1974; McGrath and Solter 1984b). The preferential inactivation of the paternal X chromosome in the extraembryonic tissues (Takagi and Sasaki 1975; West et al. 1977; Harper et al. 1982) is another striking example of the imprinting of the genome (Rastan and Cattanach 1983; Lyon and Rastan 1984). Other differences between the extraembryonic and embryonic tissues, such as differential activity of the cellular homologues of virus oncogenes (Muller et al. 1983) and the under-methylation of some dispersed and centromeric repetitive DNA sequences in the extraembryonic tissues but not in the embryo (Chapman et al. 1984), have been detected. It is not known if these differences relate to the influence of the paternal genome on the proliferation of the trophoblast.

It would be of interest to establish precisely when, during gametogenesis and especially spermatogenesis, specific imprinting of the genome is completed, and also, until what stage of embryonic development paternal and maternal genomes retain all the information necessary for development to term and if such information is lost at different times from the two genomes. It may be expected that imprinting is reversed in the primordial germ cells, perhaps at the same time when, during female gametogenesis, reactivation of the inactive X chromosome has been detected (Johnston 1981; Lyon and Rastan 1984). The process of genomic imprinting is then probably reinitiated during gametogenesis.

REFERENCES

Bagshawe, K.D. and S.D. Lawler. 1982. Unmasking moles. *Br. J. Obstet. Gynaecol.* **89**: 255.

Barton, S.C., M.A.H. Surani, and M.L. Norris. 1984. Roles of paternal and maternal genomes in mouse development. *Nature* **311**: 374.

Bensaude, O., C. Babinet, M. Morange, and F. Jacob. 1983. Heat shock proteins, first major products of zygotic gene activity in mouse embryo. *Nature* **5**: 331.

Bernstein, R.M. and B.B. Mukherjee. 1972. Control of RNA synthesis in 2-cell and 4-cell mouse embryos. *Nature* **238**: 457.

Borsuk, E. 1982. Preimplantation development of gynogenetic diploid mouse embryos. *J. Embryol. Exp. Morphol.* **69**: 215.

Chapman, V., L. Forrester, J. Sanford, N. Hastie, and J. Rossant. 1984. Cell lineage-specific undermethylation of mouse repetitive DNA. *Nature* **307**: 284.

Clegg, K.B. and L. Piko. 1982. RNA synthesis and cytoplasmic polyadenylation in the one-cell mouse embryo. *Nature* **295**: 342.

Flach, G., M.H. Johnson, P.R. Braude, R.A.S. Taylor, and V.N. Bolton. 1982. The transition from maternal to embryonic control in the 2-cell mouse embryo. *Eur. Mol. Biol. Organ. J.* **1**: 681.

Gardner, R.L. 1972. An investigation of inner cell mass and tropoblast tissues following their isolation from the mouse blastocyst. *J. Embryol. Exp. Morphol.* **28**: 279.

Graham, C.F. 1971. Virus assisted fusion of embryonic cells. *Acta Endocrinol. (Copenh. Suppl.)* **153**: 154.

——— . 1974. The production of parthenogenetic mammalian embryos and their use in biological research. *Biol. Rev. Camb. Philos. Soc.* **49**: 399.

Harper, M.I., M. Fosten, and M. Monk. 1982. Preferential paternal X inactivation in extraembryonic tissue of early mouse embryo. *J. Embryol. Exp. Morphol.* **67**: 127.

Hoppe, P.C. and K. Illmensee. 1977. Microsurgically produced homozygousdiploid uniparental mice. *Proc. Natl. Acad. Sci. U.S.A.* **74**: 5657.

——. 1982. Full term development after transplantation of parthenogenetic embryonic nuclei into fertilized mouse eggs. *Proc. Natl. Acad. Sci. U.S.A.* **79**: 1912.

Johnson, D.R. 1974. Hairpin-tail: A case of post-reductional gene action in the mouse egg? *Genetics* **76**: 795.

Johnston, P.G. 1981. X chromosomes activity in female germ cells of mice heterozygous for Searle's translocation T(X:16) 16 H. *Genet. Res.* **37**: 317.

Kaufman, M.H. 1978. The experimental production of mammalian parthenogenetic embryos. In *Method in mammalian reproduction* (ed. J.C. Daniel, Jr.), p. 21. Academic Press, New York.

Kaufman, M.H., S.C. Barton, and M.A.H. Surani. 1977. Normal postimplantation development of mouse parthenogenetic embryos to the forelimb bud stage. *Nature* **265**: 53.

Lyon, M.F. and S. Rastan. 1984. Parental source of chromosome imprinting and its relevance for X chromosome inactivation. *Differentiation* **26**: 63.

Mann, J.R. and R.H. Lovell-Badge. 1984. Inviability of parthenogenones is determined by pronuclei, not egg cytoplasm. *Nature* **310**: 66.

Markert, C.L. 1982. Parthenogenesis, homozygosity, and cloning in mammals. *J. Hered.* **73**: 390.

McGrath, J. and D. Solter. 1983. Nuclear transplantation in the mouse embryo by microsurgery and cell fusion. *Science* **220**: 1300.

——. 1984a. Completion of mouse embryogenesis requires both the maternal and paternal genomes. *Cell* **37**: 179.

——. 1984b. Maternal T^{hp} lethality in the mouse is a nuclear, not cytoplasmic, defect. *Nature* **308**: 550.

Modlinski, J.A. 1980. Preimplantation development of microsurgically obtained haploid and homozygous diploid mouse embryos and effect of pretreatment with cytochalasin B on enucleated eggs. *J. Embryol. Exp. Morphol.* **60**: 153.

Morris, J. 1968. The XO and OY chromosome constitution in the mouse. *Genet. Res.* **12**: 125.

Muller, R., I.M. Verma, and E.D. Adamson. 1983. Expression of c-*onc* genes: c-*fos* transcripts accumulate to high levels during development of mouse placenta, yolk sac and amnion. *Eur. Mol. Biol. Organ. J.* **2**: 679.

Niemierko, A. 1975. Induction of triplody in the mouse by cytochalasin B. *J. Embryol. Exp. Morphol.* **34**: 279.

Rastan, S. and B.M. Cattanach. 1983. Interaction between the Xce locus and imprinting of the paternal X chromosome in mouse yolk-sac endoderm. *Nature* **303**: 635.

Sanford, J., L. Forrester, V. Chapman, A. Chandley, and N. Hastie. 1984. Methylation patterns of repetitive DNA sequences in germ cells of Mus musculus. *Nucl. Acids Res.* **12**: 2823.

Sawicki, J.A., T. Magnuson, and C.J. Epstein. 1981. Evidence for expression of the paternal genome in the 2 cell mouse embryos. *Nature* **294**: 450.

Stevens, L.C. 1975. Comparative development of normal and pathenogenetic mouse embryos, early testicular and ovarian teratomas, and embryoid bodies. In *Teratomas and differentiation* (ed. M.I. Sherman and D. Solter), p. 17. Academic Press, New York.

———. 1978. Totipotent cells of parthenogenetic origin in a chimaeric mouse. *Nature* **276**: 266.

Stevens, L.C., D.S. Varnum, and E.M. Eicher. 1977. Viable chimaeras produced from normal and parthenogenetic mouse embryos. *Nature* **269**: 515.

Surani, M.A.H. 1984. Differential roles of paternal and maternal genomes during embryogenesis in the mouse. *BioEssays* **1**: 224.

Surani, M.A.H. and S.C. Barton. 1983. Development of gynogenetic eggs in the mouse: Implications for parthenogenetic embryos. *Science* **222**: 1034.

Surani, M.A.H., S.C. Barton, and M.H. Kaufman. 1977. Development to term of chimaeras between diploid parthenogenetic and fertilized embryos. *Nature* **270**: 601.

Surani, M.A.H., S.C. Barton, and M.L. Norris. 1984. Development of reconstituted mouse eggs suggests imprinting of the genome during gametogenesis. *Nature* **308**: 548.

Takagi, N. and M. Sasaki. 1975. Preferential inactivation of the paternally derived X chromosome in the extraembryonic membranes of the mouse. *Nature* **256**: 640.

Tarkowski, A.K. 1975. Induced parthenogenesis in the mouse. *Symp. Soc. Dev. Biol.* **33**: 107.

Van Blerkom, J. and G.O. Brockway. 1975. Quantitative patterns of protein synthesis in the preimplantation mouse embryo. I. Normal pregnancy. *Dev. Biol.* **44**: 148.

Wakasugi, N. 1974. A genetically determined incompatibility system between spermatozoa and eggs leading to embryonic death in mice. *J. Reprod. Fertil.* **41**: 85.

West, J.D., W.I. Frels, V.M. Chapman, and V.E. Papaioannou. 1977. Preferential expression of the maternally derived X chromosome in the mouse yolk sac. *Cell* **12**: 873.

Whittingham, D.G. 1980. Parthenogenesis in mammals. In *Oxford reviews of reproductive biology* (ed. C.A. Finn), vol. 2, p. 205. Clarendon Press, Oxford.

Wolgemuth, D.J. 1983. Synthetic activities of the mammalian early embryo: Molecular and genetic alterations following fertilization. In *Mechanism and control of animal fertilization* (ed. J.F. Hartmann), p. 415. Academic Press, New York.

Haploid Gene Expression and the Mouse t-Complex

KEITH WILLISON AND KEITH DUDLEY
Chester Beatty Laboratories,
Institute of Cancer Research,
Fulham Road, London SW3 6JB, England

OVERVIEW

A set of cDNA clones are described which are expressed in various of the cell types in the mouse seminiferous epithelium. Some of the cDNAs, prototype pB1.4, are accumulated during the later stages of spermatogenesis in haploid spermatids. It is not known whether this accumulation of mRNA represents continued transcription of the pB1.4 gene in spermatids. The possibility of gene transcription occurring in haploid cells is relevant to the observed genetic properties of *t*-haplotype 17th chromosomes, in particular transmission ratio distortion. This is the phenomenon whereby sperm bearing a *t*-haplotype chromosome, which have differentiated in a heterozygous t/+ testis, preferentially fertilize up to 99% of the eggs. We discuss the effects of *t*-haplotype genetic loci upon spermatogenesis and speculate on how these genes might be usefully studied by experiments with transgenic mice containing them.

INTRODUCTION

Our group is interested in studying gene expression during spermatogenesis in mice and has been analyzing a set of cDNA clones which we recently isolated (Dudley et al. 1984). Some members of the set are copies of messenger RNAs that appear to be accumulated during the later stages of spermatogenesis in haploid cells (spermatids). Indirect methods have been used to address the question of which cell types these genes are transcribed in and have involved the analysis of stable RNA isolated from the testes of mutant mice blocked in different stages of spermatogenesis and of normal mice of different ages. Although all the data points to the likelihood of gene transcription in haploid spermatids, proof is lacking because spermatogenesis occurs in a syncytium of cells, each one derived from a single diploid spermatogonium and giving rise to 256 spermatozoa (Dym and Fawcett 1971). Since it is unknown what degree of synchrony is maintained during differentiation within a syncytium, it is conceivable that a mRNA molecule could be synthesized in the nucleus of a diploid cell at an earlier developmental stage and transported through the large pores of the syncytium to the cytoplasm of the spermatid. This problem of the existence (or not) of haploid gene transcription is important in its own right. If products were to be synthesized, or not, in a haploid cell and vice versa in its meiotic partner, and if the pair of cells were derived from a diploid stem cell heterozygous for alleles involved in sperm function, then sperm of different fertilizing capacities could be produced. Clearly, this cannot be a common phenomenon or Mendelian genetic ratios would not generally be observed; but there are a few examples occurring in a number of organisms including *Drosophila* and *Mus musculus* (Crow 1979).

In male *Drosophila melanogaster* heterozygous for segregation distortion (SD) alleles (Hartl and Hiraizumi 1976) as many as half of the 64 spermatids in the germinal syncytium may develop abnormally and be resorbed. The remaining viable spermatids all contain the chromosome 2 carrying the mutant SD allele, since progeny analysis demonstrates that all eggs are fertilized by SD bearing spermatozoa. The SD effects are haploid cell specific but seem to be initiated in the diploid primary spermatocyte as shown by genetic analysis using nondisjunction for parts of chromosome 2. The molecular basis of segregation distortion is not known and the genetics is very complex, involving both SD and other responding loci and showing background effects of other chromosomes.

In mouse there is a group of loci in the t-complex of chromosome 17 which cause transmission ratio distortion (Lyon 1984). Transmission ratio distortion seems to be genetically similar to SD in *Drosophila* (Lyon 1984) but is physiologically different. Briefly, male mice heterozygous for a normal 17th chromosome and a *t*-haplotype chromosome produce an ejaculate of spermatozoa consisting of roughly equal numbers of sperm of each genotype (Silver and Olds-Clarke 1984). However the *t*-haplotype bearing spermatozoa fertilize a greater proportion of eggs, up to 99% in some combinations of particular *t*-haplotype chromosomes and genetic background. Various genetic and physiological experiments are consistent with the idea that the distortion effects are generated during spermatogenesis. This could occur in diploid stem cells such that a *t*-haplotype gene product would act upon the wild-type chromosome but not upon itself, or it could occur during the syncitial stage by a number of mechanisms involving sharing or nonsharing of RNAs or proteins or other products. The isolation and characterization of genes expressed in haploid spermatids may lead to a molecular analysis of transmission ratio distortion.

RESULTS

cDNA probes made from 18s poly(A)$^+$ RNA fraction of testes, liver, and spleen were used to screen a 1.4×10^4 event cDNA library made from poly(A)$^+$ RNA isolated from the testes of CBA/Ca mice. We defined 48 colonies which corresponded to RNAs expressed much more abundantly in the testes than in the liver or spleen. Of these 48 clones we selected ten for further analysis. Northern blots of RNA made from testis and liver were individually hybridized to probes made by nick translation of each of the ten clones. The results of this experiment showed that six of the clones did not give a detectable signal with the liver RNA although hybridizing strongly to the testis RNA; two hybridized equally to both RNAs; and two gave a weak signal with the liver RNA and a strong signal with the testes RNA.

All the RNAs detected were between 15s and 20s as expected since we screened the library with an 18s probe.

In order to determine which, if any, of the ten clones corresponded to mRNAs expressed postmeiotically in the testes, we hybridized each of the ten nick translated clones to Northern blots of RNA made from the testis of mice at different stages of sexual maturity. RNA from the testis of 2-week and 3-week-old mice was used in this experiment since, in mice, three weeks marks the completion of meiosis and the first appearance of haploid spermatids. Figure 1 shows the results for five of the clones. Clearly four of the clones correspond to mRNAs expressed more abundantly in the RNA made from the testes of 3-week-old mice.

In order to confirm this result, RNA made from the testes of testicularly feminized mice was analyzed on Northern blots. Testicularly feminized mice, genetically *Tfm*/Y, appear outwardly to be female but have small testes and no accessory glands. On histological examination, the testes of these mice have been shown to lack any germ cells past the first meiotic division. In this respect they resemble the testes of 2-week-old mice. The results of this experiment confirmed the data from the earlier Northern using RNA from 2-week and 3-week-old mice. A summary of the data obtained by Northern blotting is shown in Table 1.

By screening the testes cDNA library against probes made from liver RNA and spleen RNA, we had hoped to eliminate cDNA clones containing highly repeated sequences as well as to select sequences expressed uniquely in the testis. To determine the copy number of the ten clones that we selected for further study, we used them as nick translated probes on Southern blots. The result for two of the clones is shown in Figure 2. Clone 33 hybridizes to three restriction fragments of

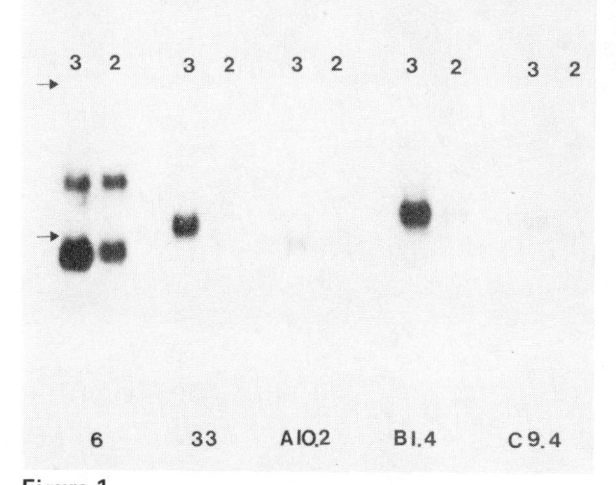

Figure 1

Northern blot analysis of cDNA clones using RNA isolated from 2-week-old and 3-week-old mice. Strips of nitrocellulose with one lane of 3-week RNA (*3*) and one lane of 2-week RNA (*2*) were hybridized to nick translated cDNA clones identified by the numbers at the bottom of the figure. The upper and lower arrows correspond to the position where 28s and 18s RNA migrate.

Table 1
Characteristics of the Ten Clones Studied

	Liver RNA	Testis RNA			Tfm RNA	Size of RNA
		Adult	2-Week	3-Week		
2	+	+	+	+	+	15s
6	weak+;+	+;+	+;+	+;+	+;+	16s, 20s
9	–	+	–	+	–	18s
14	–	+	–	+	–	15s
20	–	+	–	+	–	18s
33	–	+	very weak+	+	–	19s
A10.2	–	+	–	+	–	18s
B1.4	weak+	+	weak+	+	weak+	19s
C1	weak+	+	–	+	–	20s
C9.4	–	+	–	+	–	19s

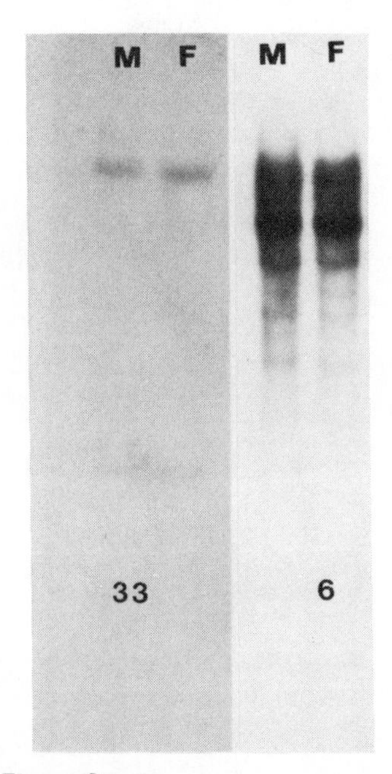

Figure 2
Southern blot analysis of two cDNA clones. DNA was made from the livers of either male (M) or female (F) CBA/Ca mice, digested with *Eco*RI and the fragments resolved on a 1% agarose gel. The Southern blot of this gel was hybridized to nick translated cDNA probes, identified by the numbers at the bottom of the figures.

*Eco*RI digested CBA/Ca liver DNA, suggesting that it corresponds to a single or low copy number gene. Clone 6, which from the Northern blot analyses (Fig. 1) hybridized to two transcripts, corresponds to a sequence only moderately repeated in the mouse genome. Of the ten clones we analyzed, six were shown to be single or low copy number and the remaining four moderately repeated; one of which (clone 2) was shown to be a transcript from the mitochondrial genome.

None of the ten clones were specific for male DNA. This result was considered a possibility since some testicular transcripts might derive from the male sex-specific Y chromosome. All of the ten clones gave an identical pattern of hybridization on Southern blots with *Eco*RI digested DNA isolated from the livers of male or female CBA/Ca mice (Fig. 2). DNA from inbred strains of mice has restriction fragment length polymorphism for pB1.4 and other of the clones, and this should allow the genetic mapping of these clones using classical genetics and recombinant inbred lines (Bailey 1971).

DISCUSSION

Analysis of Mouse Male Sterile Mutations

We have described an approach to studying gene expression in spermatogenesis which should lead to the genetic mapping and identification of specific genes. There are approximately 30 mapped genes in the mouse which affect the process of spermatogenesis to cause sterility (Green 1981). Some are specific in the sense that one particular cell type is affected and two examples are *Sl* (Chr.10) and *W* (Chr.5). Primordial germ cells (PGCs) are absent from the gonads of Sl/Sl homozygotes because although the PGCs normally migrate towards the genital ridges during development, they fail to proliferate there. This is not due to defects in the Sl/Sl PGCs but to the failure of cueing from the cells in the microenvironment. W/W^v mice lack PGCs because of intrinsic defects in stem cell populations although in fact this has only been formally proven for the progenitor cells of erythrocytes and melanocytes in this mutant. Autosomal reciprocal translocations may cause male sterility by affecting genes at the breakpoints or by moving genes closer to heterochromatin. The severity of the defects may be affected by genetic background such as H-2 linked genes (Forejt 1976).

These examples of different genetic effects on spermatogenesis highlight the problem of analyzing transmission ratio distortion caused by mutant genes in the t-complex. The phenomenon itself is extremely variable. It can depend on the time of mating relative to the time of ovulation (Braden 1958). Buehr and McLaren (1984) discuss the physiology of and possible explanations for this type of effect. There are considerable effects of genetic background particularly of the non-*t*-haplotype 17th chromosome (Bennett et al. 1983). Thus, is the phenomenon non-specific in the sense that too many gene products and activities are involved which will confound analysis or can specific loci be identified at the molecular level? Lyon

(1984) has genetically dissected the transmission ratio distortion properties of complete *t*-haplotype chromosomes into four components (t-complex distorters 1, 2, and 3 and t-complex responder). She discusses the possible identity of these loci with the t-complex polypeptide (Tcp1, Tcp3-9) loci of Silver et al. (1983). It seems that these polypeptides are the best candidates for Tcds and we are presently screening our cDNA clones to identify any which might map to the t-complex.

Use of Transgenic Mice to Study the Male Germline

If transmission ratio distortion can be reduced to a small number of genes and/or genetic elements, it may be possible to reintroduce them into mice and dissect the relative contribution of each as Lyon (1984) has begun to do by conventional genetic techniques using partial *t*-haplotype analysis. Unfortunately, although it may be possible to produce transgenic mice containing potential Tcds, interpretation may be difficult. Palmiter et al. (1984) have recently found an usual transgenic mouse MyK-103 that has a foreign DNA insert composed of two copies of a metallothionein-thymidine kinase fusion gene (pMK) oriented as inverted repeats. This female mouse produced fertile male progeny which never transmitted the pMK insert containing chromosome to their own progeny. Their proposed explanation is that the foreign DNA insert has disrupted a gene expressed during haploid stages of spermatogenesis. The identification and mapping of the pMK interrupted gene may allow its association with a known mouse sterility locus.

In summary, the availability of numerous mutations that affect male sterility in the mouse, coupled with differential cDNA cloning techniques and the development of the transgenic mouse technique, seem to make spermatogenesis an exciting area to explore with the techniques of molecular biology.

ACKNOWLEDGMENTS

This work is supported by grants from the MRC/CRC Joint Committee for the Institute of Cancer Research to K.W.

REFERENCES

Bailey, D.W. 1971. Recombinant-inbred strains, an aid to finding identity, linkage, and function of histocompatibility and other genes. *Transplantation* **11**: 325.

Bennett, D., A.K. Alton, and K. Artzt. 1983. Genetic analysis of transmission ratio distortion by t-haplotypes in the mouse. *Genet. Res.* **41**: 29.

Braden, A.W.H. 1958. Influence of time of mating on the segregation ratio of alleles at the T-locus in the house mouse. *Nature* **181**: 786.

Buehr, M. and A. McLaren. 1984. Interlitter variation in progeny of chimaeric male mice. *J. Reprod. Fertil.* **72**: 213.

Crow, J.F. 1979. Genes that violate Mendel's Rules. *Sci. Am.* **240**: 134.

Dudley, R.K., J. Potter, M.F. Lyon, and K.R. Willison. 1984. Analysis of male sterile mutations in the mouse using haploid stage expressed cDNA probes. *Nucl. Acids Res.* **12**: 4281.

Dym, M. and D.W. Fawcett. 1971. Further observations on the number of spermatogonia, spermatocytes, and spermatids connected by intercellular bridges. *Biol. Reprod.* **3**: 195.

Forejt, J. 1976. Spermatogenic failure of translocation heterozygotes affected by H-2 linked gene in mouse. *Nature* **260**: 143.

Green, M.C., ed. 1981. *Genetic variants and strains of the laboratory mouse.* Fischer, Stuttgart, New York.

Hartl, D.L. and Y. Hiraizumi. 1976. Segregation distortion. In *The genetics and biology of Drosophila* (ed. M. Ashburner and E. Novitski), vol. 16, p. 616. Academic Press, New York.

Lyon, M.F. 1984. Transmission ratio distortion in mouse t-haplotypes is due to multiple distorter genes acting on a responder locus. *Cell* **37**: 621.

Palmiter, R.D., T.M. Wilkie, H.Y. Chen, and R.L. Brinster. 1984. Transmission distortion and mosaicism in an unusual transgenic mouse pedigree. *Cell* **36**: 869.

Silver, L.M. and P. Olds-Clarke. 1984. Transmission ratio distortion of mouse t-haplotypes is not a consequence of wild-type sperm degeneration. *Dev. Biol.* **105**: 250.

Silver, L.M., J. Uman, J. Danska, and J.I. Garrels. 1983. A diversified set of testicular cell proteins specified by genes within the mouse t-complex. *Cell* **35**: 35.

Session 2: Endogenous Viruses and Viral Vectors

Instability of Ecotropic Proviruses in RF/J-derived Hybrid Mice

NANCY A. JENKINS AND NEAL G. COPELAND
Department of Microbiology and Molecular Genetics
University of Cincinnati
College of Medicine
Cincinnati, Ohio 45267

OVERVIEW

RF/J inbred mice carry three endogenous ecotropic murine leukemia virus (MuLV) proviruses, *Emv-1, Emv-16,* and *Emv-17,* but do not spontaneously express virus due to both Mendelian and non-Mendelian factors. However, hybrid mice derived from male RF/J mice and females of other inbred strains such as SWR/J express ecotropic virus. To further study the expression of the RF/J proviral loci, we derived several SWR/J lines congenic for the RF/J endogenous ecotropic MuLV genomes. During these studies we found that *Emv-16* and *Emv-17* segregated as tightly linked loci which were highly expressed in both male and female hybrid mice. The viral genotype of progeny mice derived from male carriers for *Emv-16* and *Emv-17* was limited to the two parental RF/J proviruses, but the ecotropic proviral DNA content of progeny from female carriers was often highly unstable. Amplification occurred and many new ecotropic proviral genomes were detected. The mechanism by which these new viral sequences are acquired is unclear but may involve germline virus infection. Finally, this system may be very useful for generating virally-induced mutations that are amenable to study at the molecular level.

INTRODUCTION

Endogenous retroviruses are transmitted as part of the normal genetic complement of a wide variety of animal species. In the mouse, there are several well characterized classes of endogenous retroviruses. One interesting family is the ecotropic murine leukemia viruses which are defined on the basis of their ability to infect and replicate only in mouse cells. Examination of DNA from various inbred mouse strains by Southern analysis has revealed that the number of endogenous ecotropic MuLV loci can vary from none to six copies per haploid mouse genome (Jenkins et al. 1982) and, to date, at least 40 distinct ecotropic proviral integration sites have been identified (Jenkins et al. 1982; N.G. Copeland and N.A. Jenkins, unpubl.). Several endogenous ecotropic proviral loci have been mapped to specific chromosomes and, in general, these loci are dispersed throughout the mouse genome (Ihle et al. 1979; Kozak and Rowe 1980; Jenkins et al. 1981; N.A. Jenkins and N.G. Copeland, unpubl.).

Expression of endogenous ecotropic proviruses varies greatly with inbred strain and age and appears to be under stringent cellular control. The mechanisms regulating provirus expression are unclear, but may be at the transcriptional level (Cabradilla et al. 1976; Thomson et al. 1980) and controlled by linked cis-acting cellular DNA sequences (Copeland and Cooper 1979), DNA methylation (Groudine et al. 1981; Harbers et al. 1981; Stuhlmann et al. 1981; Hoffmann et al. 1982), or both. The chromosomal integration site of a given proviral locus also seems to influence the time in development and the tissue of virus activation (Jaenisch et al. 1981). Additionally, some endogenous proviruses have been molecularly cloned and shown to carry small mutations within their genome that inhibit expression (Copeland et al. 1984a).

The distribution of various endogenous ecotropic MuLV loci follows known strain and substrain relationships, suggesting that virus integration occurred after speciation, but before the establishment of inbred strains, and that these integrations are relatively stable. In fact, by combining the number of years that several low viremic strains have been maintained independently, it was possible to estimate that no new proviruses had been gained or lost in the germline of these mice in 1000 years or 2500 generations of inbreeding (Jenkins et al. 1982). These results are in contrast to those obtained in highly viremic strains, such as AKR (Quint et al. 1982; Steffen et al. 1982), AKR/J-derived recombinant inbred strains (Jenkins et al. 1981; Steffen et al. 1982) and congenic NIH.AKV sublines of mice (Rowe and Kozak 1980). At a low frequency, these mice have been shown to acquire spontaneously new endogenous ecotropic MuLV DNA sequences. Acquisition of new ecotropic proviral loci in these strains is presumed to result from the rare infection of the germline by the virus these mice spontaneously produce throughout life (Rowe and Kozak 1980). An estimate of the rate at which new endogenous ecotropic proviruses are becoming fixed in the germline of these viremic strains is one new provirus every 15-30 years or 37-75 generations of inbreeding (Jenkins et al. 1982; Quint et al. 1982).

As part of an ongoing study in our laboratory, we have been interested in deriving congenic strains that carry independently various endogenous ecotropic proviral loci so that we could compare their expression, stability, and lymphoma incidence on a single host genetic background. SWR/J was chosen as the background strain for these studies because it does not carry any endogenous ecotropic proviral loci (Jenkins et al. 1982) and it is permissive for the expression of most nondefective endogenous ecotropic MuLV genomes (Jolicoeur 1979).

One of the several strains used to donate endogenous ecotropic MuLV proviruses was RF/J. RF/J mice carry three endogenous ecotropic proviruses that we have designated *Emv-1, Emv-16,* and *Emv-17* (endogenous ecotropic murine leukemia virus locus -1, -16, and -17; respectively). However, RF/J mice do not spontaneously express ecotropic MuLV due to both Mendelian (Pincus et al. 1971; Mayer et al. 1978) and non-Mendelian mechanisms. The non-Mendelian mechanism involves a maternal resistance factor, designated MRF, that is transferred by RF/J females to their progeny (Mayer et al. 1980). F_1 hybrid mice, derived from male but not female RF/J mice, do express ecotropic MuLV (Chen et al. 1980), and it has been shown that RF/J mice transmit a single dominant gene for ecotropic MuLV expression (Mayer et al. 1980). To study further the expression and stability of each RF/J provirus, we attempted to derive SWR/J lines congenic for *Emv-1, Emv-16,* and *Emv-17.*

RESULTS

Segregation and Expression of the Endogenous Ecotropic Proviruses of RF/J Mice

To begin the derivation of SWR/J lines congenic for each RF/J ecotropic provirus, SWR/J females were crossed to male RF/J mice and male (SWR/J × RF/J)F_1 mice were then backcrossed to SWR/J females. Spleens were surgically removed from the progeny of the first backcross (N_2) generation and high molecular weight DNA was prepared from each tissue sample. DNAs were digested to completion with *Pvu*II, an enzyme that cleaves most ecotropic proviruses twice (Rands et al. 1981), Southern blotted, and hybridized with an ecotropic virus-specific probe (Chattopadhyay et al. 1980; Jenkins et al. 1982). Under these conditions, a single 3′ cell-virus junction fragment containing 3.0 kb of viral DNA can be detected for each endogenous AKR-like ecotropic provirus present. Following this analysis, the three proviruses of RF/J mice are resolved as 4.3 kb (*Emv-1*), 4.8 kb (*Emv-16*), and 6.1 kb (*Emv-17*) viral DNA-containing fragments (Fig. 1). Southern analysis of 17 N_2

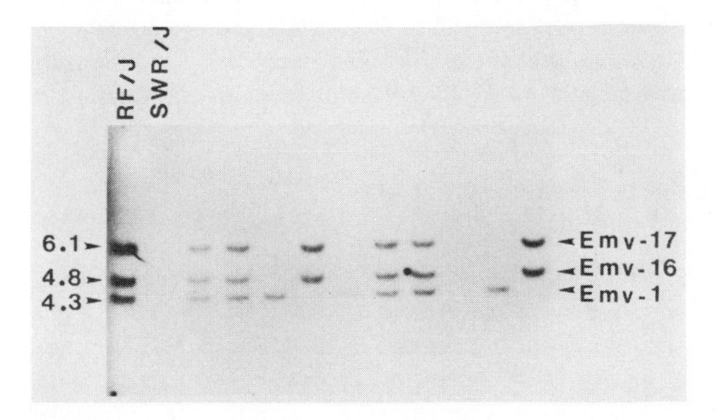

Figure 1
Segregation of the endogenous ecotropic MuLV loci of RF/J in [SWR/J × (SWR/J × RF/J)F_1] hybrid mice. The three endogenous ecotropic proviruses of RF/J mice are resolved, following *Pvu*II digestion and Southern analysis, as 4.3 kb (*Emv-1*), 4.8 kb (*Emv-16*), and 6.1 kb (*Emv-17*) viral DNA-containing fragments. SWR/J mice lack endogenous ecotropic viral DNA sequences. Southern analysis of DNAs of ten of the progeny of the first backcross generation (N_2) is shown.

generation progeny (SWR/J × (SWR/J × RF/J)F_1) indicated that five animals inherited *Emv-1*, *Emv-16*, and *Emv-17*; two animals were ecotropic virus negative; six animals inherited only *Emv-1*; and four animals inherited *Emv-16* and *Emv-17* (Fig. 1 and data not shown). These results suggested that *Emv-1* segregates independently of *Emv-16* and *Emv-17* and that *Emv-16* and *Emv-17* are linked.

To establish whether the two ecotropic proviruses were actually linked, three of the four N_2 animals that carried just *Emv-16* and *Emv-17* were backcrossed to SWR/J mice. We then examined 43 N_3 progeny by Southern analysis for the segregation of *Emv-16* and *Emv-17*, and again we were unable to separate the two proviruses. To date, 455 animals segregating for *Emv-16* and *Emv-17* from the N_2 through the N_7 generation have been analyzed and no recombinants between *Emv-16* and *Emv-17* have been detected. Therefore, at the 95% confidence level, *Emv-16* and *Emv-17* are within 0.7 cM of each other (Green 1981), assuming free recombination can occur.

The proviral locus designated *Emv-1* is also carried by a number of other inbred strains (Jenkins et al. 1982). This locus on mouse chromosome 5 is rarely expresed in young mice and has been shown by molecular cloning and DNA transfection analysis to be defective (K.W. Hutchison et al., unpubl.). This locus is also not expressed in N_2 generation SWR/J-RF/J hybrid mice and has not been pursued further.

As expected, since RF/J male mice were used in the initial crosses (thus relieving virus suppression due to the maternal resistance factor), most N_2 mice carrying *Emv-16* and *Emv-17* were viremic (H.G. Bedigian, pers. comm.). Additional analysis of both adult male and female hybrid mice from the N_3 through the N_5 generation that carried *Emv-16* and *Emv-17*, again indicated that they were highly viremic. However, since we have been unable to separate genetically these two proviral loci, we do not know if both or only one of these proviruses are expressed. Additionally, it is not known when, with respect to the age of the mouse, virus expression first begins.

Amplification of Ecotropic Proviruses in SWR/J-RF/J Hybrid Mice

In an attempt to genetically separate *Emv-16* and *Emv-17*, male as well as female mice carrying the two proviruses were backcrossed to SWR/J mice. To date, several hundred progeny from many individual male carriers (N_2 through N_7) have been examined and the ecotropic viral DNA content of the progeny was strictly limited to *Emv-16* and *Emv-17*, which were inherited in a normal Mendelian manner.

Emv-16 and *Emv-17* were also inherited in a normal Mendelian manner in the progeny of female carriers for the linked parental viruses. However, in addition to *Emv-16* and *Emv-17*, many newly acquired proviruses were observed in somatic tissues of some progeny mice that were not present in parental mice. Representative results obtained from 11 of these mice following Southern analysis of spleen DNAs are shown in Figure 2.

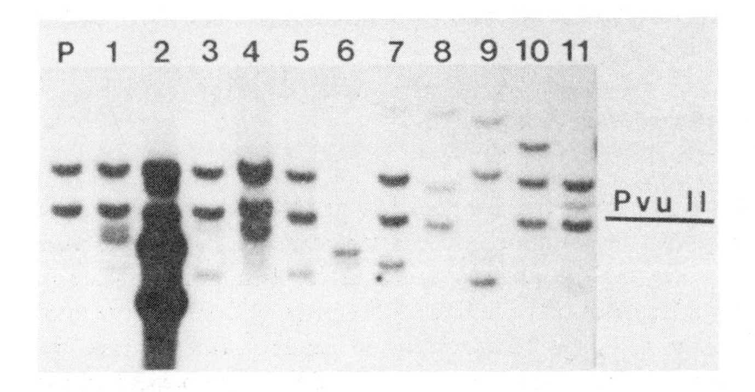

Figure 2
Amplification of the ecotropic proviral DNA content in SWR/J-RF/J hybrid mice. Southern analysis, following *Pvu*II digestion, of DNAs prepared from 11 (lanes *1-11*) progeny mice derived from female carriers (lane *p*) of *Emv-16* and *Emv-17* is shown.

Some mice appeared to carry *Emv-16* and *Emv-17* as well as a single additional ecotropic provirus (lanes *5* and *10*). Some mice carried more than one newly acquired provirus either in combination with *Emv-16* and *Emv-17* (lanes *1, 3, 4, 7, 11*) or alone (lanes *2, 6, 8,* and *9*). Furthermore, the intensity of hybridization of the newly acquired proviruses varied from DNA to DNA as well as within a single DNA sample, suggesting that these proviruses were acquired at different developmental stages.

The newly acquired provirus detected in lane *10* was approximately equal in hybridization intensity to the parental RF/J viruses which must be present at one copy each per diploid mouse genome. This suggests that the new provirus was acquired at the one-cell stage. Many of the other ecotropic MuLV genomes were much less intense, suggesting that these proviruses were acquired during early embryonic development and that the animals were mosaic. Finally, one mouse was identified (lane *2*) that had acquired multiple new proviruses, some of which appeared to be present at more than one copy per diploid mouse genome following *Pvu*II digestion. Extensive restriction analysis indicated that at least ten new ecotropic genomes were acquired and this animal did not inherit *Emv-16* and *Emv-17* (data not shown). The amplification was specific for viral DNA sequences and did not result from a loading artifact or plasmid contamination since hybridization of this DNA with several unique sequence probes gave single bands equal in size and intensity to control DNA samples (data not shown).

To date, 20 of 132 progeny generated from nine different female (N_2 through N_5) carriers for *Emv-16* and *Emv-17* have collectively acquired more than 50 new ecotropic proviruses. Restriction enzyme analysis has suggested that, in general, these proviruses are nondefective. Analysis of undigested DNA failed to detect any unintegrated viral DNA (data not shown), indicating that they are integrated within the mouse genome.

Germline Transmission of the Newly Acquired Ecotropic Proviruses

To determine if the newly acquired viral genomes detected by Southern analysis of spleen DNAs were present in the germline, all mice still available for analysis were backcrossed to SWR/J mice and the progeny analyzed for their endogenous ecotropic viral DNA content. Representative results for the progeny derived from mice analyzed in Figure 2, lanes *4, 7,* and *11* are shown in Figure 3. In each case, the newly acquired proviruses originally detected in spleen DNA were present in the germline. As expected, following germline transmission, these new proviral loci appeared to be present at one copy per cell.

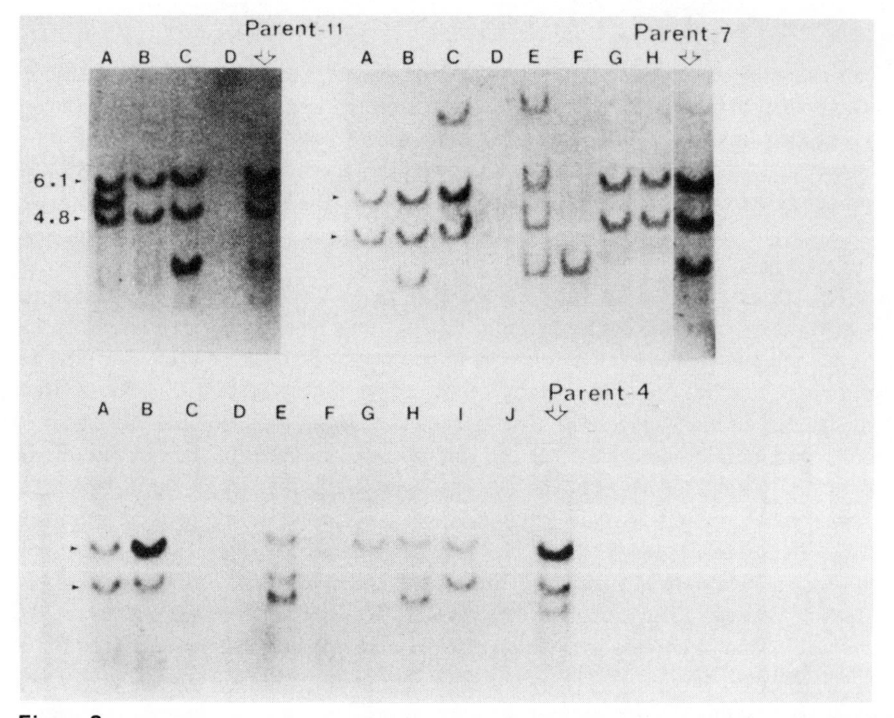

Figure 3
Germline transmission of newly acquired ecotropic proviruses. Mice carrying newly acquired proviruses analyzed in lanes *4, 7,* and *11* (Fig. 2) were backcrossed to SWR/J mice. Southern analysis, following *Pvu*II digestion, of DNAs prepared from four progeny (*A-D*) of parent 11, eight progeny (*A-H*) of parent 7, and ten progeny (*A-J*) of parent 4 is shown. The positions of the parental *Emv-16* (4.8 kb) and *Emv-17* (6.1 kb) proviruses are indicated.

Furthermore, the relative frequencies at which these new proviruses were transmitted was roughly proportional to their copy number as determined by relative hybridization intensities. For example, parent 7 (Figs. 2 and 3) carried, in addition to *Emv-16* and *Emv-17*, two new ecotropic MuLV genomes that following *Pvu*II digestion and Southern analysis were resolved as 12.5-kb and 3.5-kb viral DNA-containing fragments. Parent 7 was highly mosaic for the 12.5-kb fragment in somatic tissues and this was also reflected in the germline. Only three of 17 (18%) progeny inherited the 12.5-kb fragment whereas *Emv-16* and *Emv-17* were inherited by ten of 17 (59%) mice and the 3.5-kb fragment was transmitted to seven of 17 (41%) mice (data not shown).

Parent 4 (Figs. 2 and 3) actually acquired two new ecotropic proviruses, one of which comigrated with *Emv-17* following *Pvu*II digestion and Southern analysis. Additional restriction analysis indicated that this new provirus also comigrated with *Emv-16* and/or *Emv-17* following digestion with at least four other enzymes and could only be distinguished from the parental proviruses following digestion with *Hin*dIII (data not shown). Nevertheless, segregation analysis indicated that this provirus was unlinked and nonallelic to *Emv-16* and *Emv-17* as well as the other newly acquired proviral genome transmitted by parent 4. In fact, all newly acquired proviruses analyzed to date segregated independently in breeding studies. Therefore, these new proviruses appear to be integrated at many different sites in the mouse genome.

Currently, 16 of these newly acquired ecotropic MuLV loci are being maintained. Mice inheriting a particular locus are identified by Southern analysis of spleen DNA and are subsequently backcrossed to SWR/J mice. These new endogenous proviral loci have been designated *Sre,* SWR/J-RF/J-associated ecotropic proviral locus, 1 through 16. Also, an SWR/J line congenic for *Emv-16* and *Emv-17* is being derived using only male virus carriers. All of these lines are being used to evaluate the structure and expression of each ecotropic proviral locus as well as to determine whether any of the new viral insertions were mutagenic.

DISCUSSION

Endogenous ecotropic murine leukemia viruses are thought to originate from rare germline infection. As ecotropic MuLVs are generally noncytopathic to the host, infection of a germ cell can result in a new provirus' becoming part of the host genetic complement and the provirus is then transmitted in a normal Mendelian manner.

We have shown here that two of the three endogenous ecotropic proviruses of RF/J mice (*Emv-16* and *Emv-17*) segregate as very tightly linked loci and, to date, no recombinants between the two have been detected. Restriction enzyme analysis suggested that the two viral loci are not tandemly duplicated (data not shown). The RF/J strain was derived from the RFM/Un strain (Popp and Amos 1965) and although both strains carry *Emv-1*, only RF/J carries the two linked proviruses. Whether this linkage occurred by two chance integration events within the germline of a progenitor of the RF/J strain or whether it reflects a single integration event followed by a gene duplication and/or rearrangement event has yet to be determined. Additionally, it is likely that RF/J mice were unintentionally outcrossed to AKR mice during inbreeding (Popp and Amos 1965) and *Emv-16* and *Emv-17* may therefore be of AKR origin.

The molecular cloning of *Emv-16* and *Emv-17* and sequence analysis of flanking cellular DNA from each end of each provirus should allow us to determine whether or not both proviruses were acquired by an infection mechanism. Additional restriction analysis should resolve the orientation of the cellular DNA between the two proviruses with respect to the wild-type chromosome. If the cellular DNA between the two proviruses is inverted with respect to wild-type DNA, then recombination would be suppressed and the distance between the two proviruses may not have been accurately estimated by our breeding studies. Finally, transfection of NIH 3T3 cells with each molecular clone should allow us to determine whether one or both of the proviruses are nondefective.

Transmission of *Emv-16* and *Emv-17* via female SWR/J-RF/J mice results in the generation of many new ecotropic proviral loci. The viral DNA sequences are acquired at different developmental stages and more than one new provirus can be acquired by a single embryo. The mechanism by which the new viral sequences are acquired is unknown but most likely results from exogenous virus infection of the embryo.

Retroviruses are structurally and functionally analogous to the transposable elements of prokaryotes, yeast, and *Drosophila* (Varmus 1982). However, to date, there is no evidence for intracellular transposition events involving ecotropic MuLV proviruses. Acquisition of new ecotropic proviral loci in mice is associated with highly viremic female but not male parental mice. For example, in the derivation of NIH mice congenic for a highly expressed AKR endogenous ecotropic provirus, *Akv-1*, Rowe and Kozak (1980) detected new proviruses only in progeny derived from viremic females. Crosses of virus negative females and viremic males resulted in the stable transmission and segregation of *Akv-1*. In addition, Quint et al. (1982) identified an endogenous AKR provirus that appeared to represent a somatic recombinant, suggesting that it was acquired by exogenous virus infection.

Several explanations may account for these observations. Infectious ecotropic virus is produced in many tissues of the mouse and, in particular, in the female reproductive tract (Rowe and Pincus 1972). Additionally, it is clear that mouse embryos are susceptible to ecotropic MuLV infection, as embryos exposed to exogenous Moloney murine leukemia virus have been shown to acquire new proviral loci (Jaenisch et al. 1981). Therefore, it will be important in SWR/J-RF/J hybrid mice to determine (1) when *Emv-16* and/or *Emv-17* are first expressed, (2) which tissues express virus, and (3) at what stages in development the viral sequences are acquired. Experiments designed to address these questions are in progress.

The frequency of acquisition of new ecotropic proviral loci in certain SWR/J-RF/J hybrid mice greatly exceeded what would have been expected based upon previous reports (Jenkins et al. 1982; Quint et al. 1982; Steffen et al. 1982). However, these estimates are, at best, tenuous. Few substrains were studied and the analysis was done only after extensive inbreeding. Any provirus that was lost, rather than fixed, during inbreeding would not be scored. Studies are now in progress to construct SWR/J hybrids with other high viremic strains such as AKR/J and C58/J to determine the frequency at which these hybrids acquire new proviral loci.

The newly acquired proviruses of SWR/J-RF/J hybrid mice appear to be integrated in many different chromosomal locations. Sixteen of these loci are available for further analysis and we are currently evaluating whether any of these insertions were mutagenic. That mutating viral insertions have occurred within the mouse genome has clearly been demonstrated in two cases (Jaenisch et al. 1983; Copeland et al. 1984b). However, to date, no phenotypic dominant mutations have been observed in any of the SWR/J-RF/J hybrid mice. We are now determining whether any recessive mutations have been generated. Retroviral induced mutations are particularly amenable to experimental analysis since the viral sequences can be used to retrieve the affected gene which can then be studied in detail at the molecular level.

The instability of ecotropic proviruses in SWR/J-RF/J hybrid mice is conceptually similar to the instability of P-elements in dysgenic crosses of *Drosophila* (Engels 1983). In both cases, the mobile element must be donated by a specific parent and the cross must be done in a single direction. Insertion of P-elements into new locations in the *Drosophila* genome leads to the generation of mutations. If this is also true with respect to ecotropic proviruses in SWR/J-RF/J hybrid mice, then this system may be potentially useful for generating mutations susceptible to further experimental intervention.

ACKNOWLEDGMENTS

We thank Barbara Lee and Marie Menzietti (The Jackson Laboratory) and Debbie Swing and Charles Sabatos (University of Cincinnati) for their expert technical assistance. This work was supported by National Cancer Institute grants CA-38039 and CA-37283 as well as American Cancer Society grant MV-124.

REFERENCES

Cabradilla, C.D., K.C. Robbins, and S.A. Aaronson. 1976. Induction of mouse type-C virus by translational inhibitors: Evidence for transcriptional derepression of a specific class of endogenous virus. *Proc. Natl. Acad. Sci. U.S.A.* **73**: 4541.

Chattopadhyay, S.K., M.R. Lander, E. Rands, and D.R. Lowy. 1980. Structure of endogenous murine leukemia virus DNA in mouse genomes. *Proc. Natl. Acad. Sci. U.S.A.* **77**: 5774.

Chen, S., F.D. Struuck, M.L. Duran-Reynals, and F. Lilly. 1980. Genetic and non-genetic factors in expression of infectious murine leukemia viruses in mice of the DBA/2 X RF cross. *Cell* **21**: 849.

Copeland, N.G. and G.M. Cooper. 1979. Transfection by exogenous and endogenous murine retrovirus DNAs. *Cell* **16**: 347.

Copeland, N.G., H.G. Bedigian, C.Y. Thomas, and N.A. Jenkins. 1984a. DNAs of two molecularly cloned endogenous ecotropic proviruses are poorly infectious in DNA transfection assays. *J. Virol.* **50**: 437.

Copeland, N.G., K.W. Hutchison, and N.A. Jenkins. 1984b. Excision of the DBA ecotropic provirus in dilute coat-color revertants of mice occurs by homologous recombination involving the viral LTRs. *Cell* **33**: 379.

Engels, W.R. 1983. The P family of transposable elements in *Drosophila. Annu. Rev. Gen.* (ed. H.L. Roman et al.) **17**: 315. Annual Reviews, Palo Alto, California.

Green, E.L. 1981. *Genetics and probability in animal breeding experiments*, p. 25. Oxford University Press, New York.

Groudine, M., R. Eisenman, and H. Weintraub. 1981. Chromatin structure of endogenous retroviral genes and activation by an inhibitor of DNA methylation. *Nature* **292**: 311.

Harbers, K., A. Schnieke, H. Stuhlmann, D. Jahner, and R. Jaenisch. 1981. DNA methylation and gene expression: Endogenous retroviral genome becomes infectious after molecular cloning. *Proc. Natl. Acad. Sci. U.S.A.* **78**: 7609.

Hoffmann, J.W., D. Steffen, J. Gusella, C. Tabin, S. Bird, D. Cowing, and R.A. Weinberg. 1982. DNA methylation affecting the expression of murine leukemia proviruses. *J. Virol.* **44**: 144.

Ihle, J.N., D.R. Joseph, and J.J. Domotor. 1979. Genetic linkage of C3H/HeJ and Balb/c endogenous ecotropic C-type viruses to phosphoglucomutase-1 on chromosome 5. *Science* **204**: 71.

Jaenisch, R., D. Jahner, P. Nobis, I. Simon, J. Lohler, K. Harbers, and D. Grotkopp. 1981. Chromosomal position and activation of retroviral genomes inserted into the germline of mice. *Cell* **24**: 519.

Jaenisch, R., K. Harbers, A. Schnieke, J. Lohler, I. Chumakov, D. Jahner, D. Grotkopp, and E. Hoffmann. 1983. Germline integration of Moloney murine leukemia virus at the *Mov-13* locus leads to recessive lethal mutation and early embryonic death. *Cell* **32**: 209.

Jenkins, N.A., N.G. Copeland, B.A. Taylor, and B.K. Lee. 1981. Dilute (*d*) coat-color mutation of DBA/2J mice is associated with the site of integration of an ecotropic MuLV genome. *Nature* **293**: 370.

——. 1982. Organization, distribution, and stability of endogenous ecotropic murine leukemia virus DNA sequences in chromosomes of *Mus musculus. J. Virol.* **43**: 26.

Jolicoeur, P. 1979. The *Fv-1* gene of the mouse and its control of murine leukemia virus replication. *Curr. Top. Microbiol. Immunol.* **86**: 67.

Kozak, C.A. and W.P. Rowe. 1980. Chromosomal mapping of ecotropic and xenotropic leukemia virus-inducing loci in the mouse. In *Animal virus genetics* (ed. B. Fields and R. Jaenisch), p. 171. Academic Press, New York.

Mayer, A., M.L. Duran-Reynals and F. Lilly. 1978. *Fv-1* regulation of lymphoma and of thymic ecotropic and xenotropic MuLV expression in mice of the AKR/J × RF/J cross. *Cell* **15**: 429.

Mayer, A., F.D. Struuck, M.L. Duran-Reynals, and F. Lilly. 1980. Maternally transmitted resistance to lymphoma development in mice of reciprocal crosses of the RF/J and AKR/J strains. *Cell* **19**: 431.

Pincus, T., J. Hartley, and W.P. Rowe. 1971. A major genetic locus affecting resistance to infection with murine leukemia viruses. I. Tissue culture studies of naturally occurring viruses. *J. Exp. Med.* **133**: 1219.

Popp, D.M. and D.B. Amos. 1965. An H-2 analysis of strain RFM/Un. *Transplantation* (Baltimore) **3**: 501.

Quint, W., H. van der Putten, F. Janssen, and A. Berns. 1982. Mobility of endogenous ecotropic murine leukemia viral genomes within mouse chromosomal DNA and integration of a mink cell focus-forming virus-type recombinant provirus in the germline. *J. Virol.* **41**: 901.

Rands, E., D.R. Lowy, M.R. Lander, and S.K. Chattopadhyay. 1981. Restriction endonuclease mapping of ecotropic murine leukemia viral DNAs: Size and sequence heterogeneity of the long terminal repeat. *Virology* **108**: 445.

Rowe, W.P. and T. Pincus. 1972. Quantitative studies of naturally occurring murine leukemia virus infection of AKR mice. *J. Exp. Med.* **135**: 429.

Rowe, W.P. and C.A. Kozak. 1980. Germline reinsertions of AKR murine leukemia virus genomes in *Akv-1* congenic mice. *Proc. Natl. Acad. Sci. U.S.A.* **77**: 4871.

Steffen, D.L., B.A. Taylor, and R.A. Weinberg. 1982. Continuing germline integration of AKV proviruses during the breeding of AKR mice and derivative recombinant inbred strains. *J. Virol.* **42**: 165.

Stuhlmann, H., D. Jahner, and R. Jaenisch. 1981. Infectivity and methylation of retroviral genomes is correlated with expression in the animal. *Cell* **26**: 221.

Thomson, J.A., P.J. Laipis, G.S. Stein, J.L. Stein, M.R. Lander, and S.K. Chattopadhyay. 1980. Regulation of endogenous type C viruses: Evidence for transcriptional control of AKR viral expression. *Virology* **101**: 529.

Varmus, H.E. 1982. Form and function of retroviral proviruses. *Science* **216**: 812.

Molecular Genetic Approaches to the Study of Murine Lymphomas

NEAL G. COPELAND,* MICHAEL L. MUCENSKI,* BENJAMIN A. TAYLOR,†
AND NANCY A. JENKINS*
*Department of Microbiology and Molecular Genetics
University of Cincinnati
College of Medicine
Cincinnati, Ohio 45267
†The Jackson Laboratory
Bar Harbor, Maine 04609

OVERVIEW

Recombinant inbred (RI) mouse strains are derived from the cross of two inbred strains and represent stable segregant populations of the two parental genotypes. As such, these strains are particularly useful for gene mapping experiments. Additionally, RI strains derived from mice that differ significantly in their disease incidence may prove to be unique resources for identifying and studying genes that affect the disease process.

In particular, we have analyzed the lymphoma susceptibility of 13 AKXD RI strains derived from AKR/J and DBA/2J mice. AKR/J has a high incidence of T-cell lymphomas while DBA/2J is a low leukemic strain. Of the 13 AKXD strains analyzed, 12 showed a high incidence of lymphoma development. Yet none of the strains was as susceptible to early lymphomas as AKR/J and the lymphoma susceptibility varied considerably among the different AXKD strains. In contrast to AKR/J, tumors appeared to be predominantly of B-cell origin although both T-cell and myeloid tumors were identified. Importantly, these strains appear to differ in host genes that affect both lymphoma susceptibility and disease type and should prove invaluable for studying the molecular genetic basis of murine lymphomas.

INTRODUCTION

A number of inbred mouse strains have been developed that have a high incidence of T-cell, B-cell, or myeloid lymphomas. In general, these strains express high titers of ecotropic (viruses that are able to infect and replicate in mouse cells) murine leukemia viruses (MuLVs) early in life, which is thought to be causally related to the high lymphoma incidence of these strains. Based on our current understanding of retroviruses, it is generally thought that these viruses are oncogenic via integrating near and altering the expression of cellular oncogenes. (Nusse and Varmus 1982; Peters et al. 1983; Corcoran et al. 1984; Cuypers et al. 1984; Shen-Ong et al. 1984).

The prototype high viremic, high leukemic mouse strain is AKR. These mice develop a high incidence of T-cell lymphomas with nearly all animals developing the disease by 7-16 months of age. The high incidence of lymphomas in these mice is correlated with the expression of two endogenous ecotropic MuLV loci designated *Emv-11* and *Emv-14* (Jenkins et al. 1982). Additionally, recombinant viruses termed mink cell focus-forming (MCF) viruses have been identified in both preleukemic and leukemic thymuses of AKR/J mice (Hartley et al. 1977; Herr and Gilbert 1983, 1984). Unlike ecotropic viruses which can infect many types of cells in the mouse, MCF viruses isolated from AKR thymuses infect and replicate selectively in immature thymocytes present in the thymic cortex (Cloyd 1983). These viruses are able to accelerate leukemia when injected into newborn AKR mice (Cloyd et al. 1980; O'Donnell et al. 1981) suggesting that they are the ultimate oncogenic virus of AKR mice.

MCF viruses are not encoded directly in the AKR germline but are thought to be generated by multiple recombination events between at least three endogenous viruses (Quint et al. 1984). One recombination event appears to take place between an ecotropic virus and a xenotropic-like virus to generate a recombinant virus that contains xenotropic viral sequences within the U3 region of the viral long terminal repeat (LTR). This xenotropic-like virus appears to be encoded by a provirus present in only one copy and in the same chromosomal site in the genomes of the BALB/c, C57BL/10, AKR, and DBA strains (Quint et al. 1984). MCF viruses also carry xenotropic sequences within the gp70 region of the viral *env* (envelope) gene which are derived from a different provirus than the one used to donate the U3 LTR sequences. Both the thymotropism and leukemogenicity of these viruses have been localized to these recombinant regions (DesGroseillers et al. 1983; Celander and Haseltine 1984; DesGroseillers and Jolicoeur 1984; Lenz et al. 1984) suggesting that the generation of MCF viruses is requisite for developing T-cell lymphomas in AKR mice.

In contrast to the high leukemic mouse strains, several other strains have been generated, such as DBA, that have a very low incidence of lymphomas. Unlike AKR, DBA mice only occasionally produce ecotropic virus and, in those cases, virus expression is usually detected only late in life. Although DBA mice are permissive for replication of most endogenous ecotropic proviruses, the single endogenous ecotropic proviral locus that they carry (designated *Emv-3*) appears to carry a small mutation that inhibits its expression (Copeland et al. 1984).

By analyzing the lymphoma incidence in crosses of AKR and DBA mice, Chen and Lilly (1982) showed that DBA mice carry at least two independently segregating dominant genes which together suppress the high lymphoma phenotype of AKR mice. None of these resistance genes were linked to *Fv-1* (Friend virus restriction locus-1) or *H-2*. In an attempt to map these DBA restriction genes and to learn more about AKR leukemogenesis, we have begun to analyze the lymphoma susceptibility of 13 AKXD recombinant inbred mouse strains that were derived from crosses of AKR/J and DBA/2J mice.

RESULTS

Lymphoma Susceptibility of 13 AKXD RI Strains

Twenty female mice of each of 13 AKXD RI strains were aged to determine their lymphoma incidence. Mice were examined twice weekly for signs of illness. Moribund mice were autopsied and slides were prepared from lymphoid tissues and major organs for histological examination. In addition, pieces of affected tissue (spleen, liver, thymus, lymph nodes) were frozen for DNA extraction. Brain tissue was frozen as a control tissue since it is usually not infiltrated with leukemic cells. Mice surviving 18 months were autopsied and examined in the same way as moribund mice. Inevitably some loss of information resulted from mice dying without autopsies, from accidental deaths, or from autopsied mice that did not reveal any obvious pathology.

Among the 13 AKXD RI strains analyzed, lymphomas were the major cause of death (Table 1). The exception was AKXD-20 where only two mice developed lymphomas; 16 mice were apparently healthy at the end of the 18 months. However, this was not surprising since AKXD-20 animals, unlike mice from the other 12 AKXD strains shown in Table 1, did not inherit any endogenous ecotropic proviral loci (Jenkins et al. 1981). Among the other 12 strains, lymphoma susceptibility varied considerably as measured by incidence and age of onset of lymphomas. These strains have been ordered according to their mean age of onset of lymphomas

Table 1
Lymphoma Susceptibility of AKXD RI Strains

AKXD RI strain	Number of lymphomas	Mean age of onset (days)	Tumor type			
			T-cell	B-cell	Myeloid	Mixed
17	14	285 ± 17	7	7	0	0
6	15	309 ± 27	7	5	1	2
14	16	342 ± 16	6	10	0	0
27	14	345 ± 31	0	13	1	0
18	16	360 ± 27	3	12	1	0
13	15	378 ± 13	3	11	1	0
3	16	416 ± 25	3	13	0	0
22	10	439 ± 39	2	8	0	0
23	13	445 ± 16	0	4	8	1
9	17	448 ± 19	3	13	1	0
15	16	471 ± 25	3	11	1	1
7	17	482 ± 18	1	13	2	1
20	2	480 ± 72	0	2	0	0
Total			38	122	16	5
Percent			21.0	67.4	8.8	2.8

in Table 1 which ranged from 285 ± 17 days (AKXD-17) to 482 ± 18 days (AKXD-7). Of the 12 high lymphoma strains only AKXD-17 had a lymphoma incidence that approached that of AKR/J mice. These results would be expected for strains that have segregated for several loci that affect lymphoma incidence.

The lymphomas identified in AKXD mice were classified by gross pathological and histological data as to T-cell, B-cell, or myeloid cell origin using the recently proposed classification system of Pattengale and Taylor (1981). The results are summarized in Table 1. According to this classification scheme, lymphoid lymphomas are categorized into five major groups based upon cell morphology, which include lymphoblastic lymphoma, small lymphocytic lymphoma, folicle center cell lymphoma, immunoblastic lymphoma, and plasma cell lymphoma. Lymphoblastic lymphomas represent a heterogeneous group of lymphomas and can be either T-cell or B-cell in origin (Pattengale and Frith 1983). However, these lymphoblastic lymphomas can tentatively be classified as T-cell or B-cell in origin depending on whether or not evidence for anterior mediastinal involvement is found, mediastinal involvement denoting a lymphoma that is of T-cell origin (Pattengale and Frith 1983). Lymphomas of the other four major morphological types are generally thought to be of B-cell origin. For simplicity, the data shown in Table 1 does not discriminate between different morphological subtypes.

In contrast to AKR/J lymphomas, the majority of lymphomas that developed in the AKXD strains (67.4%) appeared to be of B-cell origin (Table 1). Only 21.0% of the lymphomas were classified as T-cell in origin. Of 181 tumors examined, 16 (8.8%) were myeloid and five tumors (2.8%) appeared to be of mixed cell type.

T-cell lymphomas were more abundant in strains with the earliest age of onset of lymphomas such as AKXD-17, AKXD-6, and AKXD-14 (Table 1). Similar results have been reported by Zijlstra et al. (1984) who showed that T-cell lymphomas generally appear earlier than B-cell lymphomas induced by MuLVs in B10 ($H\text{-}2^b$) and B10.A ($H\text{-}2^a$) mice.

Strains such as AKXD-27 and AKXD-7 appear to die primarily of B-cell lymphomas while AKXD-23 mice showed a preponderance of myeloid lymphomas (Table 1). These results suggest that all of the genetic information necessary to specify lymphomas of any given cell type are carried by AKR/J and DBA/2J mice and that by fixing this genetic information in various different combinations in the AKXD strains, it is possible to generate new recombinant strains that are highly susceptible to lymphomas of a variety of cell types.

Immunoglobulin Gene Rearrangements in Lymphomas of AKXD Mice

Sequential rearrangement of heavy and light chain immunoglobulin gene loci occur during normal B-lymphocyte differentiation and serve as useful markers for the B-cell lineage. To confirm that the AKXD tumors assigned as B-cell using the classification scheme of Pattengale and Taylor (1981) (Table 1) were actually of B-cell origin, we decided to determine if immunoglobulin gene rearrangements could be detected in the AKXD tumors. DNAs from various AKXD lymphomatous tissues were isolated and digested with appropriate restriction enzymes. Southern analysis was performed using a ^{32}P-labeled probe representative of the J_H-region of the heavy chain (IgH) gene locus or to the J_K-region of the kappa light chain (IgK) gene locus (Tonegawa 1983). The results of these experiments are summarized in Table 2.

Of the 127 independent AKXD tumors analyzed for rearrangements within the IgH region, 108 (85.0%) showed evidence for immunoglobulin heavy chain gene rearrangement (Table 2). However, rearrangements within the IgH locus are not by themselves diagnostic for B-cells. For example, rearrangements within the IgH locus have been detected in a number of primary murine T-cell lymphomas as well as T lymphoma cell lines (Cory et al. 1980; Forster et al. 1980; Kurosawa et al. 1981; Herr et al. 1983). Further analysis of these rearrangements has demonstrated that they are aberrant rearrangements involving only D-J joinings rather than the normal V-D-J joinings (Kurosawa et al. 1981). Whether or not these D-J rearrangements have any functional significance is unknown. However, D segments have been shown to carry their own 5′ transcriptional promoter element and Reth and Alt (1984) have recently shown that transcription of rearranged heavy chain D and J segments in B-lymphoid cells can occur and lead to the production of Dμ messenger

Table 2
Immunoglobulin Gene Rearrangements in Lymphomas of AKXD RI Mice

AKXD RI strain	Immunoglobulin gene rearrangements (Fraction positive lymphomas)	
	IgH	IgK
17	5/9	0/9
6	10/13	1/13
14	10/10	8/10
27	12/12	4/12
18	9/10	6/10
13	14/16	12/16
3	12/13	7/13
22	6/7	2/7
23	3/9	1/9
9	11/11	1/11
15	6/7	6/7
7	10/10	5/10
20	ND	ND
Total positive	108	53
Percent positive	85.0	41.7

IgH = Immunoglobulin heavy chain genes; IgK = immunoglobulin kappa light chain genes

RNA that can be translated into a short Dμ protein with a variable D-J *N*-terminus. Rearrangements within the heavy chain locus have also been reported to occur within cells of the myeloid lineage although they appear to occur infrequently (Rovigatti et al. 1984; M.L. Mucenski et al., in prep.). Therefore, although the IgH rearrangements detected in the AKXD tumors do not by themselves allow their classification as B-cell tumors, these rearrangements do confirm that they are monoclonal tumors of lymphoid origin.

Rearrangements within the kappa light chain locus seem to be much more diagnostic for cells of the B lineage. For example, we have consistently failed to detect rearrangements within the IgK region in a large number of lymphomas obtained from strains that get T-cell lymphomas (AKR/J, C58/J, and HRS/J) or myeloid lymphomas (BXH-2) (M.L. Mucenski et al., in prep.). In contrast, of 127 AKXD lymphomas analyzed, rearrangements within the IgK region were seen in 53 (41.7%) cases.

As expected, strains such as AKXD-17 and AKXD-6 which, by histological analysis, appeared susceptible to T-cell lymphomas, showed few if any rearrangements within the IgK locus (Table 2). Additionally, lymphomas from AKXD-23 mice, which have a high incidence of myeloid tumors, only showed evidence for IgK rearrangement in one of nine lymphomas analyzed. One notable exception was strain AKXD-14. These animals appeared about equally susceptible to T-cell and B-cell lymphomas (Table 1) although most of the lymphomas analyzed carried rearrangements within the IgK region (Table 2). Some of the IgK rearrangements occurred in AKXD-14 lymphomas that had been tentatively classified as T-cell in origin (Table 1), indicating that these tumors were misclassified. These results serve to emphasize the need to confirm the histological typing by the use of other cell lineage markers.

Failure to detect rearrangements within the IgK region does not automatically mean that a lymphoma is not of B-cell origin as B-cell lymphomas at very early stages of differentiation have been identified that have rearranged their IgH but not IgK genes (Alt et al. 1981).

As discussed previously, some lymphomas that show evidence of IgH but not IgK rearrangement may be of T-cell origin. Recently, DNA clones representative of the α- and β-chain genes encoding the T-cell receptor have been identified (Chien et al. 1984; Hedrick et al. 1984; Malissen et al. 1984; Saito et al. 1984). Like the IgH and IgK genes, these genes also rearrange during T-cell differentiation and, furthermore, these rearrangements seem to be specific for T-lymphocytes. Therefore, these rearrangements may provide useful markers for cells of the T lineage, and we are currently screening the AKXD lymphomas with probes representative of these genes.

Somatically-acquired Proviruses in Lymphomas of AKXD Mice

T-cell lymphomas of AKR/J mice are associated with recombinant MCF viruses. Unlike ecotropic viruses, MCF viruses can efficiently infect and replicate in cells of the thymic cortex and it is thought that these viruses may induce T-cell lymphomas by integrating near, and affecting the expression of, cellular oncogenes that are involved in regulating normal lymphoid differentiation. The predominant class of somatic proviruses detected in lymphoma DNAs from other high T-cell lymphoma strains such as C58/J and HRS/J are also of the MCF type (M.L. Mucenski et al., in prep.), further suggesting that MCF viruses are intimately involved in generating T-cell lymphomas. In support of this model, Herr and Gilbert (1983) have shown that thymic lymphomas of AKR/J mice contain multiple (in some cases over 15) somatically acquired MCF proviruses. Very few ecotropic proviruses were detected.

In contrast to AKR/J mice, the predominant type of lymphoma seen in AKXD mice appears to be of B-cell origin. Therefore, it was of interest to determine if the B-cell lymphomas in AKXD mice are associated with MCF proviruses. High-molecular-weight DNAs were prepared from brain tissue of each lymphomatous animal. As brain is usually not highly infiltrated with tumor tissue, its analysis serves to discriminate proviral sequences that are present in the germline of the different AKXD mice. DNAs were also isolated from two tissues of each animal that were highly infiltrated with tumor cells. By analyzing two lymphomatous tissues from each animal it was possible to distinguish viral sequences that were acquired early in the disease process and, at the same time, determine if the tumors were monoclonal in origin.

DNAs were digested to completion with *Pvu*II and Southern blotted with a ^{32}P-labeled probe that was specific for ecotropic proviruses (Chattopadhyay et al. 1980). Importantly, *Pvu*II cleaves each ecotropic provirus twice to generate a single 3′ cell-virus junction fragment detectable with the ecotropic virus-specific probe (Jenkins et al. 1982). In subsequent experiments, DNAs were also analyzed, following *Pvu*II digestion, with a hybridization probe that recognizes both ecotropic as well as the oncogenic type MCF proviruses identified in AKR/J mice (Herr and Gilbert 1983). This hybridization probe is derived from sequences within the p15E region of the viral *env* gene and hybridizes to the same 3′ cell-virus junction fragment detected with the ecotropic virus-specific probe. Since the MCF probe hybridizes to both ecotropic and MCF proviruses, MCF proviruses are identified as those proviruses that hybridize with the MCF, but not the ecotropic, virus probe.

The results of this analysis are summarized in Table 3. It appears that 85-90% of the lymphomas are monoclonal and contain somatic proviruses. Tumors that did not carry somatic proviruses were not analyzed further and are not included in the results in Table 3. Unlike AKR/J lymphomas, most of the somatic proviruses detected in the AKXD lymphomas (72.7% of the lymphomas analyzed) were ecotropic proviruses. Only six of 132 lymphomas analyzed contained exclusively MCF provirus whereas 30 lymphomas (22.7%) contained both ecotropic and MCF pro-

Table 3
Somatically-acquired Proviruses Detected in Lymphomas of AKXD RI Mice

AKXD RI strain	*Rmcf* allele	Somatically-acquired proviruses (Fraction positive lymphomas)		
		Ecotropic	MCF	Both
17	S	0/9	2/9	7/9
6	S	3/13	2/13	8/13
14	S	13/13	0/13	0/13
27	S	10/13	1/13	2/13
18	R	5/10	0/10	5/10
13	R	14/16	0/16	2/16
3	R	13/13	0/13	0/13
22	R	6/7	1/7	0/7
23	S	8/9	0/9	1/9
9	R	9/11	0/11	2/11
15	R	8/8	0/8	0/8
7	R	7/10	0/10	3/10
Total positive		96	6	30
Percent positive		72.7	4.5	22.7

S = Susceptible allele; R = resistant allele

viruses. Of all the AKXD strains analyzed, two strains, AKXD-17 and AKXD-6, had the highest percentage of lymphomas containing MCF proviruses. Interestingly, these strains also appear to be the most susceptible to T-cell lymphomas (Tables 1 and 2).

Recently, a new gene on mouse chromosome 5 (*Rmcf*) has been identified that determines whether or not cells can easily be infected with MCF virus (Hartley et al. 1983). As expected, AKR/J mice carry the susceptible (S) allele. Importantly, DBA/2J mice carry the resistant (R) allele which has therefore segregated within the different AKXD strains. The genotypes of the AKXD strain with respect to *Rmcf* have been determined by Hartley et al. (1983) and are summarized in Table 3.

The average age of onset of lymphomas in the five AKXD strains shown in Table 3 that carry the permissive *Rmcf* allele was 345 ± 27 days compared to 404 ± 17 days for the seven strains that inherited the nonpermissive *Rmcf* allele, suggesting that mice carrying the permissive allele die slightly earlier from lymphomas. This might be expected if MCF viruses induce T-cell lymphomas which occur slightly earlier than B-cell lymphomas (Zijlstra et al. 1984). However, no clear correlation between the *Rmcf* genotype and MCF viruses was seen (Table 3). For example, AKXD-14 mice carry the susceptible *Rmcf* allele yet only ecotropic proviruses were detected in AKXD-14 lymphomas. Also, several lymphomas containing MCF proviruses were identified in strains carrying the resistant allele (e.g., AKXD-18). These results suggest that other genes, perhaps endogenous proviral genomes involved in the generation of MCF viruses, have segregated during the generation of the AKXD strains.

DISCUSSION

We found 12 of 13 AKXD strains aged to be highly susceptible to lymphomas. The one resistant strain, AKXD-20, did not inherit any endogenous ecotropic proviral loci, which probably accounts for its low lymphoma phenotype. Among the 12 susceptible strains, the lymphoma incidence varied considerably as did the lymphoma type. T-cell, B-cell, and myeloid lymphomas were identified with the majority of lymphomas being of B-cell origin. The shift from predominantly T-cell lymphomas seen in the parental AKR/J strain was accompanied by a corresponding shift in the type of somatic virus identified in the lymphomas. Whereas AKR/J lymphomas contain primarily MCF proviruses, the AKXD lymphomas contained mostly ecotropic proviruses.

From the type of Southern analysis described in Table 3, it was not possible to determine if the ecotropic proviruses detected in the AKXD lymphomas were themselves recombinant viruses. AKR/J ecotropic proviruses are only weakly oncogenic and it is thought that a recombination event within the U3 region of the LTR is necessary to generate a highly oncogenic ecotropic virus (DesGroseillers et al. 1983; Celander and Haseltine 1984; DesGroseillers and Jolicoeur 1984; Lenz et al. 1984). Recombinant viruses of this type would not be discriminated with the hybridization probes used in Table 3.

The U3 region of the viral LTR has been shown to contain enhancer-like sequences suggesting that lymphomas of different cell types may be generated by specific recombinant viruses carrying enhancers that are able to function in one cell type or another. This hypothesis is being examined by molecular cloning somatic proviruses from AKXD tumors of different cell types so that their structure, particularly within the U3 region of the LTR, can be deduced at the nucleotide sequence level.

Another ultimate goal of these experiments is to identify host genes which have segregated during the inbreeding of the AKXD strains that have an effect on some aspect of the disease process. One such class of host genes may include endogenous proviral sequences that are involved in generating different oncogenic recombinant viruses. In addition, other genes, such as the *Rmcf* gene described previously, may be identified which affect the age of onset of disease or the lymphoma type. Recombinant inbred strains are ideally suited for these experiments as they represent stable segregant populations of the two parental genotypes. All data collected is cumulative. As new genes are identified, they can be analyzed to determine their effect, if any, on the disease process. Furthermore, as more AKXD strains are analyzed, the chance of establishing linkage relationships increases. Currently, we are aging another 12 independent AKXD RI strains. Data collected from these strains will be combined with the data that we have already obtained and will be useful for verifying the significance of any tentative lineage relationships.

Finally, we would like to determine if the tumors identified in AKXD mice are generated via the integration of proviruses near cellular oncogenes. To date, we have analyzed the AKXD tumors for viral integrations near two cellular oncogenes, *c-myc* and *Pim-1* since viral integrations near these oncogenes have been found to occur in some murine lymphomas (Corcoran et al. 1984; Cuypers et al. 1984). Rearrangements of the sort expected for viral integrations were detected near *c-myc* in ten of 153 (6.5%) and near *Pim-1* in seven of 132 (5.3%) of the lymphomas analyzed (M.L. Mucenski et al., in prep.). Most of the *c-myc* integrations occurred in tumors that did not have viral integrations near *Pim-1*, although two AKXD-27 lymphomas were identified that appeared to have viral integrations near both *c-myc* as well as *Pim-1*. Thus, in 11–12% of the tumors, rearrangements were detected in the vicinity of these two oncogenes. From these results, it is not clear whether these rearrangements are of primary or secondary importance in lymphomagenesis in these strains although their presence suggests that they have some role in the disease process.

If viral integrations play a primary role in inducing lymphomas in these strains, then clearly viral integrations must be occurring near other cellular oncogenes besides *c-myc* and *Pim-1*. AKXD RI strains seem to be well suited for identifying these cellular genes. Unlike AKR/J tumors that contain multiple somatic MCF proviruses, AKXD lymphomas carry few somatic proviruses, averaging only 2–4 proviruses per tumor. In addition, many tumors have been identified that appear to carry a single somatically acquired provirus. If viral integrations near cellular oncogenes represent primary events in lymphomagenesis then the presence of multiple independent tumors that carry a single exogenously acquired provirus should prove invaluable for identifying these cellular genes. Currently, we are molecularly cloning several of these proviruses and their flanking cellular sequences in an attempt to identify these cellular genes.

ACKNOWLEDGMENTS

We thank Michelle Higgins for excellent technical assistance. This work was supported by grant MV-124 (N.A.J. and N.G.C.) from the American Cancer Society and Public Health Service grants CA-37283 (N.G.C. and N.A.J.) and CA-33093 (B.A.T.) from the National Cancer Institute. The Jackson Laboratory is fully accredited by the American Association for Accreditation of Laboratory Animal Care.

REFERENCES

Alt, F., N. Rosenberg, S. Lewis, E. Thomas, and D. Baltimore. 1981. Organization and reorganization of immunoglobulin genes in A-MuLV-transformed cells: Rearrangement of heavy but not light chain genes. *Cell* 27: 381.

Celander, D. and W.A. Haseltine. 1984. Tissue-specific transcription preference as a determinant of cell tropism and leukaemogenic potential of murine retroviruses. *Nature* **312**: 159.

Chattopadhyay, S.K., M.R. Lander, E. Rands, and D.R. Lowy. 1980. The structure of endogenous murine leukemia virus DNA in mouse genomes. *Proc. Natl. Acad. Sci. U.S.A.* **77**: 5774.

Chen, S. and F. Lilly. 1982. Suppression of spontaneous lymphoma by previously undiscovered dominant genes in high- and low-incidence mouse strains. *Virology* **118**: 76.

Chien, Y.-h., D.M. Becker, T. Lindsten, M. Okamura, D.I. Cohen, and M.M. Davis. 1984. A third type of murine T-cell receptor gene. *Nature* **312**: 31.

Cloyd, M.W. 1983. Characterization of target cells for MCF viruses in AKR mice. *Cell* **32**: 217.

Cloyd, M.W., J.W. Hartley, and W.P. Rowe. 1980. Lymphomagenicity of recombinant mink cell focus-inducing murine leukemia viruses. *J. Exp. Med.* **151**: 542.

Copeland, N.G., H.G. Bedigian, C.Y. Thomas, and N.A. Jenkins. 1984. DNAs of two molecularly cloned endogenous ecotropic proviruses are poorly infectious in DNA transfection assays. *J. Virol.* **49**: 437.

Corcoran, L.M., J.M. Adams, A.R. Dunn, and S. Cory. 1984. Murine T lymphomas in which the cellular *myc* oncogene has been activated by retroviral insertion. *Cell* **37**: 113.

Cory, S., J.M. Adams, and D.J. Kemp. 1980. Somatic rearrangements forming active immunoglobulin μ genes in B and T lymphoid cell lines. *Proc. Natl. Acad. Sci. U.S.A.* **77**: 4943.

Cuypers, H.T., G. Selten, W. Quint, M. Zijlstra, E.R. Maandag, W. Boelens, P. van Wezenbeek, C. Melief, and A. Berns. 1984. Murine leukemia virus-induced T-cell lymphomagenesis: Integration of proviruses in a distinct chromosomal region. *Cell* **37**: 141.

DesGroseillers, L. and P. Jolicoeur. 1984. The tandem direct repeats within the long terminal repeat of murine leukemia viruses are the primary determinant of their leukemogenic potential. *J. Virol.* **52**: 945.

DesGroseillers, L., E. Rassart, and P. Jolicoeur. 1983. Thymotropism of murine leukemia virus is conferred by its long terminal repeat. *Proc. Natl. Acad. Sci. U.S.A.* **80**: 4203.

Forster, A., M. Hobart, H. Hengartner, and T.H. Rabbitts. 1980. An immunoglobulin heavy-chain gene is altered in two T-cell clones. *Nature* **286**: 897.

Hartley, J.W., R.A. Yetter, and H.C. Morse III. 1983. A mouse gene on chromosome 5 that restricts infectivity of mink cell focus-forming recombinant murine leukemia viruses. *J. Exp. Med.* **158**: 16.

Hartley, J.W., N.K. Wolford, L.J. Old, and W.P. Rowe. 1977. A new class of murine leukemia virus associated with development of spontaneous lymphomas. *Proc. Natl. Acad. Sci. U.S.A.* **74**: 789.

Hedrick, S.M., D.I. Cohen, E.A. Nielsen, and M.M. Davis. 1984. Isolation of cDNA clones encoding T cell-specific membrane-associated proteins. *Nature* **308**: 149.

Herr, W. and W. Gilbert. 1983. Somatically acquired recombinant murine leukemia proviruses in thymic leukemias of AKR/J mice. *J. Virol.* **46**: 70.

——. 1984. Free and integrated recombinant murine leukemia virus DNAs appear in preleukemic thymuses of AKR/J mice. *J. Virol.* **50**: 155.

Herr, W., A.P. Perlmutter, and W. Gilbert. 1983. Monoclonal AKR/J thymic leukemias contain multiple J_H immunoglobulin gene rearrangements. *Proc. Natl. Acad. Sci. U.S.A.* **80**: 7433.

Jenkins, N.A., N.G. Copeland, B.A. Taylor, and B.K. Lee. 1981. Dilute (*d*) coat colour mutation of DBA/2J mice is associated with the site of integration of an ecotropic MuLV genome. *Nature* **293**: 370.

——. 1982. Organization, distribution, and stability of endogenous ecotropic murine leukemia virus DNA sequences in chromosomes of *Mus musculus*. *J. Virol.* **43**: 26.

Kurosawa, Y., H. von Boehmer, W. Haas, H. Sakano, A. Trauneker, and S. Tonegawa. 1981. Identification of D segments of immunoglobulin heavy-chain genes and their rearrangement in T lymphocytes. *Nature* **290**: 565.

Lenz, J., D. Celander, R.L. Crowther, R. Patarca, D.W. Perkins, and W.A. Haseltine. 1984. Determination of the leukaemogenicity of a murine retrovirus by sequences within the long terminal repeat. *Nature* **308**: 467.

Malissen, M., K. Minard, S. Mjolsneŝs, M. Kronenberg, J. Goverman, T. Hunkapiller, M.B. Prystowsky, Y. Yoshikai, F. Fitch, T.W. Mak, and L. Hood. 1984. Mouse T cell antigen receptor: Structure and organization of constant and joining gene segments encoding the β polypeptide. *Cell* **37**: 1101.

Nusse, R. and H.E. Varmus. 1982. Many tumors induced by the mouse mammary tumor virus contain a provirus integrated in the same region of the host genome. *Cell* **31**: 99.

O'Donnell, P.V., E. Stockert, Y. Obata, and L.J. Old. 1981. Leukemogenic properties of AKR dualtropic (MCF) viruses: Amplification of murine leukemia virus-related antigens on thymocytes and acceleration of leukemia development of AKR mice. *Virology* **112**: 548.

Pattengale, P.K. and C.R. Taylor, 1981. Immunomorphologic classification of murine lymphomas and related leukemias. In *Proceedings of the rodent lymphoma workshop*, p. 22. National Center Toxicol. Res. Press, Jefferson, Ark.

Pattengale, P.K. and C.H. Frith. 1983. Immunomorphologic classification of spontaneous lymphoid cell neoplasms occurring in female BALB/c mice. *J. Natl. Cancer Inst.* **70**: 169.

Peters, G., S. Brooks, R. Smith, and C. Dickson. 1983. Tumorigenesis by mouse mammary tumor virus: Evidence for a common region for provirus integration in mammary tumors. *Cell* **33**: 369.

Quint, W., W. Boelens, P. van Wezenbeek, T. Cuypers, E.R. Maandanq, G. Selten, and A. Berns. 1984. Generation of AKR mink cell focus-forming viruses: A conserved single-copy xenotrope-like provirus provides recombinant long terminal repeat sequences. *J. Virol.* **50**: 432.

Reth, M.G. and F.W. Alt. 1984. Novel immunoglobulin heavy chains are produced from DJ_H gene segments in lymphoid cells. *Nature* **312**: 418.

Rovigatti, U., J. Mirro, G. Kitchingman, G. Dahl, J. Ochs, S. Murphy, and S. Stass. 1984. Heavy chain immunoglobulin gene rearrangement in acute nonlymphocytic leukemia. *Blood* **63**: 1023.

Saito, H., D.M. Kranz, Y. Takagaki, A.C. Hayday, H.N. Eisen, and S. Tonegawa. 1984. A third rearranged and expressed gene in a clone of cytotoxic T lymphocytes. *Nature* **312**: 36.

Shen-Ong, G.L.C., M. Potter, J.F. Mushinski, S. Lavu, and E.P. Reddy. 1984. Activation of the *c-myb* locus by viral insertional mutagenesis in plasmacytoid lymphosarcomas. *Science* **226**: 1077.

Tonegawa, S. 1983. Somatic generation of antibody diversity. *Nature* **302**: 575.

Zijlstra, M., R.E.Y. DeGoede, H. Schoenmakers, T. Radaszkiewicz, and C.J.M. Melief. 1984. Ecotropic and dualtropic mink cell focus-inducing murine leukemia viruses can induce a wide spectrum of H-2 controlled lymphoma types. *Virology* **138**: 198.

EK Cell Contribution to Chimeric Mice: From Tissue Culture to Sperm

MARTIN EVANS, ALLAN BRADLEY, AND ELIZABETH ROBERTSON
Department of Genetics
University of Cambridge
Cambridge CB2 3EH, England

OVERVIEW

Pluripotential stem cells grown in tissue culture directly from explanted mouse blastocysts (EK cells) may be maintained with an unaltered developmental capacity and karyotype. When reintroduced into mouse blastocysts, they take part in the development of the resulting mouse and are able to contribute to all tissues. This chimera formation is efficient; no abnormalities have been observed in the chimeric individuals; and the introduced tissue-cultured EK cells may contribute to the germline. Male (XY) EK cells in combination with a female host blastocyst are able to convert the phenotypic sex of the resulting mouse to male, and 100% of the functional sperm are derived from the EK cells. In the absence of this sex conversion, the EK cells contribute to a low proportion of the sperm, but this may be increased by using host blastocysts of genotypes that are partially fertile.

These results demonstrate an efficient route between cells in culture and the laboratory mouse genome. EK cells may be selected in culture following modification by DNA transformation or transfection. Alternatively, they may be selected for specific mutations prior to reintroduction into the embryo.

INTRODUCTION

There is a population of cells present in the early mouse embryo, which is capable of self-renewal and which provides the precursors of all cell types. These cells are, therefore, multipotential stem cells and were first discovered as the stem cells of mouse teratocarcinomas. These tumors are readily formed from early embryos and the stem cells from them may be isolated and maintained in clonally derived tissue cultures (the teratocarcinoma stem cells are termed embryonal carcinoma cells or EC cells). The homology of these cells with cells of early embryos was confirmed by the ability of certain EC cell lines to form chimeric mice when they were introduced into the cavity of 3.5-day mouse blastocysts (Brinster 1974; Mintz and Illmensee 1975; Papaioannou et al. 1975). The success of this manipulation raised the hope that EC cells might provide a route between in vitro cell culture manipulation and the complete normal mouse, but in practice two problems arose. In the first place, although a number of EC cell lines were shown to be capable of forming chimeras, most contributed to only a limited number of tissues and many proved to be further abnormal as the resulting mice developed tumors. These were either teratocarcinomas, similar to those from the parent cell line, or (usually of later onset) tumors of differentiated cell types (Papaioannou et al. 1978). Second, a contribution to functional gametes by the EC cells is needed to allow transmission of in vitro introduced mutations to the mouse gene pool, but the only EC cell line for which this has been reported is METT-1 (Stewart and Mintz 1981). Other EC cell lines that have been tested have either proved to be poor at chimera formation or to be aneuploid (Papaioannou and Rossant 1983).

We have shown that the multipotential stem cells from the early mouse embryo may be isolated directly into tissue culture and we have called them EK cells (Evans and Kaufman 1981). These cells grow directly from explanted blastocysts and represent primary cultures. Under optimal culture conditions, these cells may be maintained through many passages in an unaltered state, retaining both a normal karyotype and an unaltered propensity for differentiation (Robertson et al. 1983a). These cells differentiate very readily in vitro or as a tumor in vivo and, as will be further reported here, in chimeric combination with a developing embryo.

RESULTS

The EK cells used have all been isolated and maintained as described previously (Evans and Kaufman 1981; Robertson et al. 1983a). They were reintroduced into the cavity of 3.5-day expanded blastocysts as single cells after dissociation from tissue culture by trypsin treatment. During the course of these experiments, progressively larger numbers of EK cells (up to 15) were injected into individual blastocysts as it became apparent that this was not deleterious. The injected blastocysts were reintroduced into the uteri of pseudopregnant foster mothers and allowed to develop to term. Chimerism was assessed visually by coat color markers and in internal tissues by assay of glucose phosphate isomerase isoenzymes. We have tested only a proportion of the male chimeric mice which have been constructed using XY EK cells by breeding to assess the functional germ cell contribution from the EK cells.

Chimeric mice have been made from 17 different EK cell lines derived from different individual embryos (Table 1). Although the efficiency varies between experiments it is clear that the average percentage of chimeric animals recovered is very high (34%). Both XY and XX EK cells derived from fertilized embryos and XX EK cells derived from parthenogenetic embryos have been used. In the case of XX EK cell lines from either origin, we have observed an instability of the XX constitution so that XO and XX^{del} carrying cells arise within these lines (Robertson et al. 1983b; E. Robertson, unpubl.). It is notable that extensive chimerism is also produced from the homozygous diploid cell lines derived from haploid parthenogenetic embryos.

Table 1
Chimera Forming Efficiencies of a Variety of EK Cell Lines[c]

Cell line	Strain	Chromosomes	Injected	Born	Chimeras (%)
B2B2	129	40 XY	66	40 (60.1)	14 (35.0)
CP2	129	40 XY	123	94 (76.4)	18 (19.1)
CP3	129	40 XY	156	111 (71.1)	43 (38.7)
CP4-2	129	40 XX:XX^{del}	29	8 (27.6)	5 (62.5)
A13	129	40 XX	160	109 (68.1)	36 (33.0)
X1	Rb163H	38 XY	79	52 (73.5)	4 (7.7)
CL6	LT	40 XY	161	103 (64.0)	21 (20.4)
Rm53	Rm	– XY	63	48 (76.2)	8 (16.7)
CZ4	CFLP	– XY	8	6 (75.0)	2 (33.3)
CP1	129	40 XY	246	161 (65.7)	70 (43.4)
CC1.1	129^{cc}	40 XY	83	61 (73.5)	37 (60.6)
CC1.2	129^{cc}	40 XY	321	205 (64.0)	127 (61.8)
HD1[b]	129	40 XX^{del}	42	30 (71.4)	8 (26.6)
HD10[b]	129	40 –	113	81 (71.1)	10 (12.3)
HD14[b]	129	40 XX	110	68 (61.8)	17 (25.0)
HD15[b]	F_1[a]	– –	46	31 (67.4)	1 (3.2)
DP3[b]	F_1[a]	– –	13	10 (76.9)	4 (40.0)
			1819	1218 (67.0)	420 (34.5)

[a]F_1 (CBA × C57BL)
[b]Lines derived from parthenogenetically activated embryos (Kaufman et al. 1983)
[c]Chimeras were scored at birth or shortly afterwards by the presence of the pigmented cells derived from the EK cells in the coat or eye on the albino background of the MF1 or CFLP strains. Host embryos used with the albino Rm5.3 and CZ4 cell lines were from MF1 or CFLP females crossed to an F_1 male. Resulting chimeras were scored at birth by observation of an albino component on the pigmented background of the host embryo.

The chimerism produced by these EK cells is extensive and fine-grained as would be expected from an early incorporation into the embryoblast cell pool. Table 2 shows an analysis (using GPI isoenzyme separation) of chimerism in the tissues of 34 different mice from seven different cell lines. In many cases all tissues analyzed show a contribution from the EK cells.

To date, all the EK cell lines tested appear to proliferate and differentiate in a completely normal manner within the embryonic environment; the introduction of EK cells into embryos does not result in detectable fetal loss; additionally, none of the 420 chimeras described in Table 1 have shown tumor formation.

Following these results, we have concentrated on using XY EK cell lines with the object of establishing conditions for the reproducible achievement of germline chimerism. To date, three independently isolated lines have been shown to form functional sperm, and a total of 21 chimeric mice have transmitted an EK cell genome.

Table 2
EK Contributions to Various Tissues in Chimeras as Assessed by GPI Isoenzyme Analysis

Number	1	2	3	4	5	6	7	8	9	10	11	12	13	14	15	16	17
Liver	*	*	*	*	*	*	*	*	*	*	*	*	*	*	*	*	*
Spleen	*	*	*	*	*	*	*	*	*	*	*	*	*	*	*	*	*
Heart	*	*	*	*	*	*	*	*	*	*	*	*	*	*	*	*	*
Lung	*	0	*	*	*	*	*	*	*	0	*	–	*	*	*	*	*
Right kidney	*	*	*	*	*	*	*	*	*	*	*	*	*	*	*	*	*
Left kidney	*	*	*	*	*	*	*	*	*	*	*	*	*	*	*	*	*
Adrenals	*	–	*	*	*	*	*	*	*	*	*	0/*	*	*	*	*	*
Pancreas	*	–	–	*	*	–	*	*	*	–	*	–	–	*	*	*	*
Stomach	–	–	–	*	*	–	*	*	*	–	*	–	–	0	*	*	*
Right gonad	*	0	*	*	*	*	*	–	*	*	*	–	*	*	*	*	*
Left gonad	*	–	*	*	*	*	*	*	*	*	*	–	*	*	–	*	*
Bladder	–	–	–	*	*	*	*	*	*	*	*	*	*	0	*	*	*
Uterus	*	–	–	*	*	–	*	*	*	*	*	–	–	*	*	*	*
Muscle	*	–	*	*	*	*	*	*	*	*	*	*	*	*	*	*	*
Abdominal wall	–	–	–	*	*	–	*	*	*	*	*	–	–	*	*	*	*
Brain	*	*	*	*	*	*	*	*	*	*	*	0	*	*	*	*	*
Gut	–	*	–	*	*	*	*	*	*	*	*	*	*	*	*	*	*
Thymus	–	–	–	–	–	–	–	–	–	–	–	–	–	–	*	*	*
Tail	–	0	–	*	*	–	*	*	*	–	*	*	*	0	*	*	*
Tongue	–	–	–	*	*	–	*	*	*	–	*	–	–	0	*	*	*

* = Chimeric; 0 = nonchimeric; – = not tested

1–3, B2B2 females; 4 and 6, CL6 females; 5, CL6 male; 7, Rm5.3 female; 8 and 9, Rm5.3 males; 10, 12, and 13, A13 males; 11, A13 female; 14–16, CP1 females; 17, CP1 male (germline chimera); 18–23, CC1.2 female; 24, WCC1.2 female; 25–28, CC1.1 females; 29, CC1.1 male (germline chimera); 30, WCC1.1 female; 31, HD10 female; 32, HD14 female; 33 and 34 HD14 males

Table 2 *(continued)*

Number	18	19	20	21	22	23	24	25	26	27	28	29	30	31	32	33	34
Liver	0	0	0	*	0	*	*	*	*	*	*	*	*	*	*	*	–
Spleen	0	0	0	*	0	*	*	*	*	*	*	0	*	*	–	*	–
Heart	*	0	0	*	0	*	*	*	*	*	*	*	*	*	*	*	*
Lung	0	0	0	*	*	*	*	*	*	*	*	*	*	*	*	*	*
Right kidney	*	0	*	*	*	*	*	*	*	*	*	*	*	*	*	*	*
Left kidney	*	0	0	*	0	*	*	*	*	*	*	*	*	*	*	*	*
Adrenals	*	0	0	*	*	*	*	*	*	*	*	*	*	*	–	*	*
Pancreas	*	0	*	*	0	*	*	*	*	*	*	*	*	*	–	–	–
Stomach	0	0	0	*	0	*	*	*	*	*	*	0	*	*	–	–	–
Right gonad	*	0	0	*	–	*	*	*	*	*	*	0	*	*	–	*	–
Left gonad	0	0	0	*	–	–	*	*	*	*	*	*	*	*	–	*	–
Bladder	0	0	0	*	0	*	*	*	*	*	*	*	*	*	*	–	*
Uterus	0	0	0	*	0	*	*	*	*	*	*	*	*	*	*	*	–
Muscle	*	*	0	*	*	*	*	*	*	*	*	0	*	*	*	0	*
Abdominal wall	*	*	0	*	0	*	*	*	*	*	*	0	*	*	*	*	–
Brain	*	0	*	*	*	*	*	*	*	*	*	*	*	*	*	*	*
Gut	0	0	0	*	0	*	*	*	*	*	*	*	*	*	*	*	–
Thymus	*	0	0	*	0	*	*	*	*	*	*	0	*	–	–	–	–
Tail	0	*	0	*	0	*	*	*	*	*	*	0	*	*	*	–	*
Tongue	0	0	0	*	0	*	*	*	*	*	*	*	*	*	–	–	–

* = Chimeric; 0 = nonchimeric; – = not tested

1–3, B2B2 females; 4 and 6, CL6 females; 5, CL6 male; 7, Rm5.3 female; 8 and 9, Rm5.3 males; 10, 12, and 13, A13 males; 11, A13 female; 14–16, CP1 females; 17, CP1 male (germline chimera); 18–23, CC1.2 female; 24, WCC1.2 female; 25–28, CC1.1 females; 29, CC1.1 male (germline chimera); 30, WCC1.1 female; 31, HD10 female; 32, HD14 female; 33 and 34 HD14 males

There are three significant features that result in efficient formation of germline chimeras using XY EK cells.

1. A high proportion (> 50%) of the live born animals are overtly chimeric (Tables 1 and 3).
2. The contribution by the introduced cells to the animal is extensive (Table 2).
3. The sex ratio of the chimeric mice is distorted (Table 3).

This sex distortion and the observation of phenotypic hermaphrodites is to be expected with an inappropriate combination of the genetic sex of the EK cells and the host blastocyst. The host embryo is of either sex. Combination of XY EK cells with an XY blastocyst is expected to give rise to a male mouse. On the other hand, combination of male EK cells with a female embryo may have three possible outcomes. With low levels of chimerism, there may be little contribution of the male cells to the sex-determining accessory tissues of the genital ridge and a phenotypically female mouse will result. A higher level of contribution by the XY EK cell progeny to the cells of the genital ridge results in the male-determining factors dominating sexual differentiation, and the mouse will develop as a phenotypic and functional male. Intermediate states may give rise to an intersex development.

Table 4 shows the breeding record for some of the chimeric mice which have proved to transmit the EK-derived genome, i.e., to be germline chimeras; and Figure 1 shows such a mouse together with its family. It is noticeable that there are two clear classes of chimera; those where 100% of the functional sperm are EK cell-derived and those in which EK-derived sperm form only a small percentage of the total (1-3%). We propose that the former represent mice resulting from the combination of the male EK cells with a female embryo where the XX blastocyst-derived germ cells are unable to develop in the testicular environment, while the latter results from the XY-XY combination.

This conversion of a female embryo and 100% transmission of the EK-derived germ cells is a fortunate and convenient method for efficient use of XY EK cell lines, but is there an alternative way to depress the germ cell contribution from the host blastocyst which would possibly also be applicable to XX cell lines? We are investigating this possibility by making use of alleles at the dominant white spotting

Table 3
Sex Ratio Distortion Rates of Construction of Germline Chimeras

Cell lines	Chimeric	Chimeras		ND	Males		Germline chimeric
		Female	Male		Setup	Bred	
CP1	70	27	40	3	24	17	5
CC1.1	37	13	24	0	13	11	3
CC1.2	127	29	97	1	41	19	7
	234	69	161	4	78	47	15

Table 4
Breeding Data from 15 Germline Chimeric Mice

Male chimera	Litters	Offspring	Albino	Black agouti	Transmission (%)
CP1.3	10	89	0	89	100
CP1.5	23	251	244	7	2.8
CP1.4	18	228	227	1	0.4
CP1.11	6	93	91	2	2.1
CP1.34	6	23	0	23	100
CC1.1.5	7	105	103	2	1.9
CC1.1.3	7	76	0	76	100
CC1.1.24	5	68	67	1	1.5
CC1.2.6	9	102	0	102	100
CC1.2.8	6	80	0	80	100
CC1.2.32	10	138	137	1	0.7
CC1.2.33	8	127	126	1	0.8
CC1.2.40	7	101	100	1	1.0
CC1.2.80	6	65	0	65	100
CC1.2.67	5	55	0	55	100

Figure 1
Chimeric male mouse formed from EK cell line CC1.1 injected into a blastocyst from an albino mouse stock pictured together with its mate (a pure-breeding albino female) and the litter it has sired. All the offspring are visibly derived from paternal gametes carrying the wild-type allele at the albino locus.

locus of the mouse, the *W* locus (Silvers 1979). Alleles at this locus have effects on coat pigmentation, cause macrocytic anemia and reduce fertility. This is explained on the basis of a depression of cell lineages which are involved in migration during embryogenesis. We have used three different alleles at the *W* locus, W^{sh}, W^{v} and W^{e} which, in the homozygous condition, affect viability and fertility to different degrees. In the heterozygous condition, they exhibit characteristic dominant coat color phenotypes. We have used a variety of crosses to generate host blastocysts which possess different combinations of the *W* alleles. Table 5 presents results obtained following injection of EK cells (wild-type at the *W* locus) and carrying the distinctive *GPI-1*^{c} allele into *W*-bearing host blastocysts. Chimeric individuals were scored by GPI analysis of blood samples and males caged with females (wild-type at the *W* locus). Germline chimeric individuals may be scored by transmission of the EK-derived *GPI-1*^{c} allele in their progeny. Partial germline chimeras allow the genotype of the host blastocyst to be determined as they segregate three different alleles at the *W* locus.

The data presented in Tables 5 and 6 illustrate the potential use of the *W* locus for maximizing the incorporation of the EK cells into the germline. It is clear that the proportion of male chimeras possessing an EK-derived germline component is elevated when *W*-bearing host embryos are used: five out of seven males which were test bred (70%) are germline chimeras. This compares favorably with the results presented in Table 3 where approximately 30% of the males proved to be germline chimeric. In addition, the partially transmitting germline chimeras on a *W*-bearing blastocyst background show a significantly higher transmission of the culture-derived genome (4–75%).

DISCUSSION

The facility to be able to isolate mouse embryonic stem cells, maintain a normal cell phenotype through prolonged tissue culture, and reintroduce them into a developing embryo and hence to the laboratory mouse gene pool has now been established (Bradley et al. 1984). This technique will allow selection and modification

Table 5
Rate of Construction of Chimeras Using Host Embryos Carrying *W* Alleles

						♂ Chimeras		
Cross	Injected	Born	Chimeras[a]	♀	♂	Setup	Bred	Germ-line
$W^{e}/+ \times W^{e}/+$	53	33	13	3	10	5	2	2
$W^{v}/+ \times W^{e}/+$	24	18	6	3	3	4	3	1
$W^{v}/+ \times W^{sh}/W^{sh}$	31	17	5	3	2	2	2	2
	108	68 (63%)	24 (35%)	9	15	11	7	5

[a]Chimeras made by injecting CC1.2 cells into host blastocysts and scored by GPI assay of a blood sample from the live born animals

Table 6
Breeding Data from the Germline Chimeras Described in Table 5[b]

Germline chimera	Host embryo genotype[a]	Progeny host embryo derived	Progeny EK derived	Transmission (%)
W/CC1.2.1	$-/-$	29	29	100.0
W/CC1.2.5	$W^e/+$	24	1	4.0
W/CC1.2.T8.1	W^v/W^{sh}	5	15	75.0
W/CC1.2.T17.1	W^v/W^{sh}	19	3	13.6
W/CC1.2.T17.3	W^v/W^{sh}	6	2	25.0

[a]Host embryo genotypes established by scoring alleles segregated by the germline chimeras.
[b]The coat color phenotype of the chimeras in these experiments were very variable and impossible to predict, accordingly chimeras were assessed by the presence of GPI-1^c1^c blood (contributed by cells derived from the introduced CC1.2 cells). Chimeras were test bred by crossing to 129 females which were wild-type at the *W* locus and homozygous GPI-1^a. The detection of the GPI-1^c allele in progeny was required to demonstrate that W/CC1.2.1 and W/CC1.2.5 were germline chimeras. The other germline chimeras transmit two distinguishable *W* alleles which are dominant; wild-type progeny must therefore be culture derived. GPI analysis confirms this.

in vitro of the mouse genome and should greatly facilitate both developmental and genetic studies. Although this approach was proposed using tumor-derived EC cell lines, it has not proved generally practicable until the advent of embryo-derived EK cell lines. In order for this technique to be used routinely, it will be necessary to maintain the EK cells in an unchanged multipotential state through a variety of modification and selection procedures carried out in vitro. We are currently modifying these cells by direct DNA transformation or infection with retroviral vectors as a means towards establishing a route to transgenesis. Our preliminary results indicate that clones of cells which have been transformed and selected using the dominant selection system of the bacterial *neo* gene and the antibiotic G418 are fully competent in chimera formation.

ACKNOWLEDGMENTS

We would like to thank Pam Fletcher and Lesley Cooke for their excellent technical assistance, the Cancer Research Campaign and the Medical Research Council for their generous support of this work. A.B. is a Beit Memorial Research Fellow.

REFERENCES

Bradley, A., M. Evans, M.H. Kaufman, and E. Robertson. 1984. Formation of functional germ line chimeras from embryo-derived teratocarcinoma cell lines. *Nature* **309**: 255.

Brinster, R.L. 1974. The effect of cells transferred into the mouse blastocyst on subsequent development. *J. Exp. Med* **140**: 1049.

Evans, M.J. and M.H. Kaufman. 1981. The establishment in culture of pluripotential cells from mouse embryos. *Nature* **292**: 154.

Kaufman, M.H., E.J. Robertson, A.H. Handyside, and M.J. Evans. 1983. Establishment of pluripotential cell lines from haploid mouse embryos. *J. Embryol. Exp. Morphol.* **73**: 249.

Mintz, B. and K. Illmensee. 1975. Normal genetically mosaic mice produced from malignant teratocarcinoma cells. *Proc. Natl. Acad. Sci. U.S.A.* **72**: 3585.

Papaioannou, V.E. and J. Rossant. 1983. Effects of the embryonic environment on proliferation and differentiation of embryonal carcinoma cells. In *Cancer surveys*, vol 2, no. 1, p. 165. Oxford University Press, Oxford.

Papaioannou, V.E., M.W. McBurney, R.L. Gardner, and M.J. Evans. 1975. Fate of teratocarcinoma cells injected into early mouse embryos. *Nature* **258**: 70.

Papaioannou, V.E., R.L. Gardner, M.W. McBurney, C. Babinet, and M.J. Evans. 1978. Participation of cultured teratocarcinoma cells in mouse embryogenesis. *J. Embryol. Exp. Morphol.* **44**: 93.

Robertson, E.J., M.H. Kaufman, A. Bradley, and M.J. Evans. 1983a. Isolation properties and karyotype analysis of pluripotential (EK) cell lines from normal and parthenogenetic embryos. In *Teratocarcinoma stem cells* (ed. Martin et al., vol. 10, p. 647. Cold Spring Harbor Laboratory, Cold Spring Harbor, New York.

Robertson, E.J., M.J. Evans, and M.H. Kaufman. 1983b. X-Chromosome instability in pluripotential stem cell lines derived from parthenogenetic embryos. *J. Embryol. Exp. Morphol.* **74**: 297.

Silvers, W.K. 1979. *The coat colors of mice.* Springer-Verlag, New York.

Stewart, T.A. and B. Mintz. 1981. Successive generations of mice produced from an established culture line of euploid teratocarcinoma cells. *Proc. Natl. Acad. Sci. U.S.A.* **78**: 6314.

Expressing Foreign Genes in Stem Cells and Mice

ERWIN WAGNER,* ULRICH RÜTHER,* ROLF MÜLLER,* COLIN STEWART,*
ELI GILBOA,† AND GORDON KELLER**
*European Molecular Biology Laboratory
Postfach 10.2209
D-6900 Heidelberg, Federal Republic of Germany
†Department of Molecular Biology
Princeton University
Princeton, New Jersey 08544
**Basel Institute of Immunology
487 Grenzacherstrasse
CH-4005 Basel, Switzerland

OVERVIEW

Recombinant genes have been expressed in both murine embryonal carcinoma (EC) cells and hemopoietic cells after infection with retroviral vectors carrying the selectable neomycin resistance marker.

EC cells carrying and expressing the vector were reintroduced into mice by aggregating the cells with preimplantation embryos. A number of adult chimeras were analyzed and expression of the retroviral sequences was found in a number of chimeric tissues. Similarly, infected bone marrow cells were injected into lethally irradiated syngeneic adult mice and efficient expression of the vector was found in the hemopoietic cells of the reconstituted animal.

As a first step to studying the effect of expression of an exogenous proto-oncogene in mice, various c-*fos* gene constructs were transferred into EC stem cells. The expression of exogenous c-*fos* protein was accompanied by the appearance of differentiated cells suggesting an involvement of the c-*fos* gene in cellular differentiation.

INTRODUCTION

The introduction of recombinant genes into fertilized mouse eggs and various murine stem cells offers a powerful tool to studying the regulation of gene expression. This approach exposes the introduced gene to all possible stimuli that would occur during embryonic development and the differentiation of a particular cell lineage. Thus, the correct expression of an exogenous gene, both quantitatively and qualitatively, in the appropriate tissue should allow elucidation of the factors governing gene expression.

The transfer of certain genes, such as *onc* genes, into embryos should also help to unravel the function of their encoded gene products. In addition, such an approach could provide us with information concerning the induction of neoplastic transformation in vivo.

We are currently investigating two stem cell systems, embryonal carcinoma cells and hemopoietic stem cells as targets for genetic manipulation. Both of these stem cell types offer the advantage that they can be studied and manipulated in vitro and can then be reintroduced into the mouse for study in vivo. Our current approach has been to introduce genes into these stem cells using the advantages of retroviral vectors.

In this article we summarize results obtained in three series of experiments: (1) The isolation of embryonal carcinoma (EC) cell clones carrying retroviral vectors and the expression of these vectors in EC embryo chimeras; (2) the expression of retroviral vectors in murine hemopoietic cells, both in vitro and in vivo; and (3) the transfer of a *c-onc* gene into mouse EC cells to study the effect of its expression on cellular differentiation.

RESULTS AND DISCUSSION

Gene Transfer into EC Stem Cells and Mice by Retrovirus Infection

Embryonal carcinoma cells, the stem cells of teratocarcinomas, can lose their malignant phenotype and participate in development to form all somatic as well as germ cells after reintroduction into early embryos (Graham 1977; Mintz and Fleischman 1981; Bradley et al. 1984). Therefore, these cells offer a unique experimental system to introduce specific recombinant genes into mice (Stewart 1984). All methods that are available to manipulate the genome of eucaryotic cells can be applied to cultured EC cells. The genetically altered EC cells can be examined in undifferentiated and differentiated states prior to chimera formation, thereby providing an advantage over other methods used for introducing foreign genes into mice.

A large number of established EC cell lines, mostly derived from tumors of grafted embryos are available, and recently stem cell lines derived directly from embryos grown in vitro have been isolated (Evans and Kaufman 1981; Martin 1981; Axelrod 1984). Some of these lines, referred to as EK lines, have been reported to colonize the germline at a high frequency (Bradley et al. 1984). However, the efficiency of introducing genes into EC or EK cells by DNA-mediated gene transfer is very low (Wagner and Mintz 1982; Nicolas and Berg 1983; Stewart 1984). Recently the infectious properties of retroviruses have been exploited so that they can function as vectors for introducing exogenous genes into various cells (Mulligan 1983). A difficulty with using such vectors on EC cells is that retroviral gene expression is repressed even though infection results in the integration of proviral DNA into the host cell genome (Stewart et al. 1982; Gautsch and Wilson 1983; Niwa et al. 1983). The restriction of viral gene expression in EC cells may partly be due to the long terminal repeat's (LTR) acting as an inefficient promoter (Linney et al. 1984), although additional blocks may also exist.

Two different sets of retroviral vectors have been used to introduce and express genes in EC cells following virus infection. The first set of viral vectors consisted of viral regulatory elements (promoters, splice sites, poly A signals) to express selectable (*neo* or *gpt*) and nonselectable (human β-interferon cDNA) genes in EC cells and mice. The simplest of these vectors contained only the two LTRs and the selectable *neo* gene at the *gag* position of the proviral DNA (Fig. 1; rGag-neo). Viral stocks from various constructs were generated on NIH3T3 cells with titers of 10^6 neo-CFU/ml and were used to infect various EC cell lines. Overall, a frequency of 10^{-3} (0.1%) *neo*^{R} clones was obtained on some EC cells under selectable conditions. The clones usually contained a single and intact proviral DNA copy of the recombinant genome (Fig. 2A) and stably expressed the foreign gene in the absence of selection (E. Wagner et al., in prep.; C. Stewart et al., in prep.). A large reduction in the level of RNA expression was found in nonpermissive EC cells compared to the permissive NIH3T3 cells, which could partly account for the low frequency in obtaining *neo*^{R} colonies (Fig. 2B).

The second set of vectors contained the *neo* gene linked to the thymidine kinase (TK) promoter region and a viral *onc* gene (v-*myc*) under the control of the viral LTR (Fig. 1; MMCV-neo) (Wagner et al. 1985). Of the various EC cells infected with these viruses, which express the *neo* gene from the internal promoter, 10–100% gave rise to *neo*^{R} colonies following selection in G418-containing medium (Wagner et al. 1985). The *neo* gene was expressed in both the nonpermissive EC and the permissive NIH3T3 cells at similar levels (Fig. 2B) and transcription of *neo*-specific RNA was initiated at the proper site in the TK promoter. No *myc* expression was detected in the EC cells. Thus, genes inserted into retrovirus constructs can easily be introduced in established EC cell lines by virus infection and they can be efficiently expressed from internal promoters.

Some of these viruses containing the *neo* gene (rGag-neo and MMCV-neo) were used to infect feeder dependent pluri- and totipotent EC and EK cells, which were known to form chimeras. In one series of experiments, P10 EC cells were infected with rGag-neo and one selected clone (P10-neo1) was isolated and characterized (C.

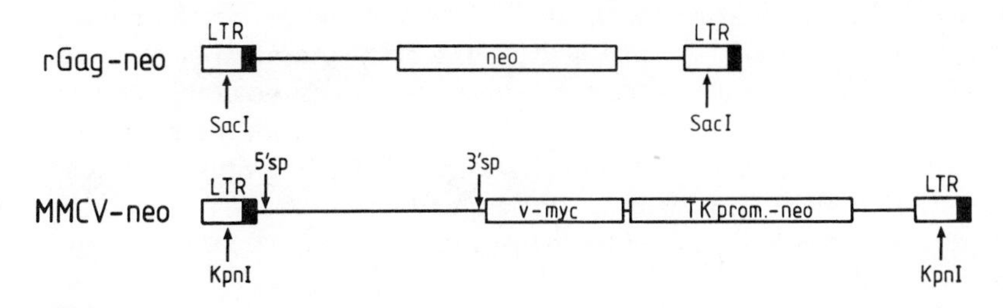

Figure 1
Structure of the retroviral vectors. The long terminal repeats (*LTRs*) and adjacent viral sequences are derived from Moloney murine leukemia virus. (*5'sp*) 5' splice donor site; (*3'sp*) splice acceptor site.

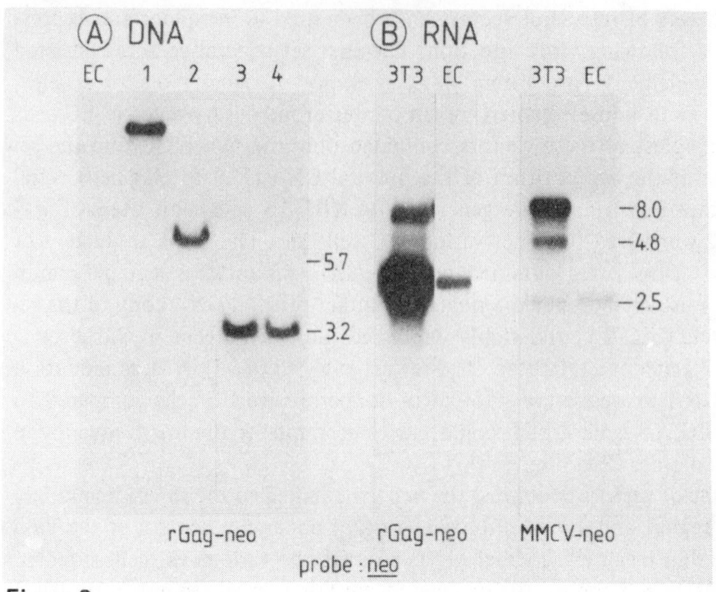

Figure 2
Analysis of DNA and RNA from EC and NIH3T3 cells infected with rGag-neo or MMCV-neo and selected in G418-containing medium. (*A*) Southern transfer analysis of genomic DNA from individual P19 clones digested with *Hin*dIII (noncutting enzyme for vector DNA; lanes *1* and *2*) and *Sac*I (cuts in the LTRs; lanes *3* and *4*); EC is uninfected P19 DNA. (*B*) RNA analysis of virus infected and selected NIH3T3 and EC cells. Expression of virus encoded *neo* and v-*myc* RNAs in pools of neo^R clones of NIH3T3 and F9 (EC) cells infected with rGag-neo or MMCV-neo (Wagner et al. 1985).

Stewart and E. Wagner, in prep.). The developmental capability of these cells was tested by reintroducing them into normal mouse embryos by the aggregation technique (Stewart 1982; Fuji and Martin 1983). At present, 34 mice have been born following manipulation and 11 of these show overt chimerism in the coat. These chimeras are currently being analyzed to establish the extent of chimerism in various tissues and to study whether expression of the introduced gene has been maintained during differentiation. In some individuals a clear correlation between the presence of the P10-neo1 cells in various organs and the expression of the *neo* gene in these tissues was found (Fig. 3). In a second series of experiments, totipotent Cp1 EK cells were infected with helper-independent MMCV-neo virus and several selected clones were isolated and used to form aggregation chimeras. The expression of the introduced genes (the *neo* and the v-*myc* gene) in different organs is currently being investigated. It thus appears that it is possible to select for the desired, genetically altered EC cell clones in vitro, and that these manipulated cells are still able to undergo differentiation and development in the embryo. It is still necessary, however, to demonstrate that this method offers an alternative to introducing genes

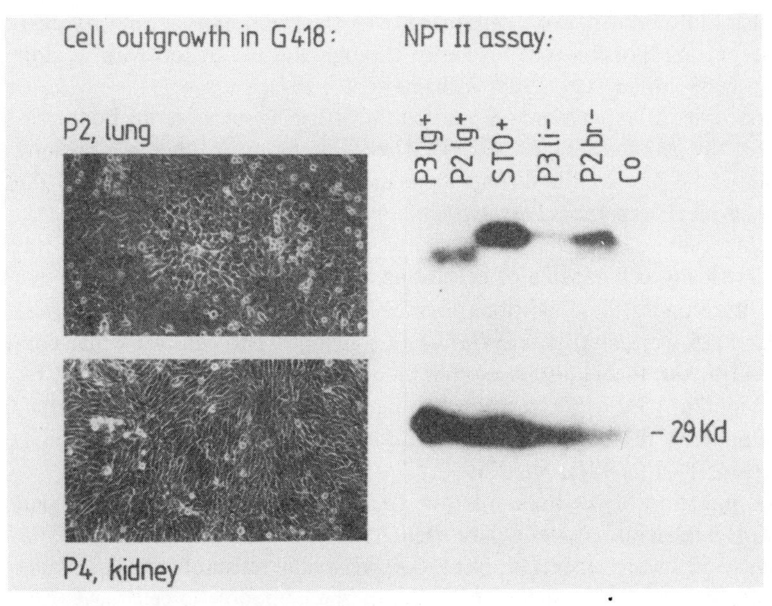

Figure 3
Expression of the introduced *neo* gene in chimeric mice obtained by embryo aggregation with an infected and preselected EC cell clone (P10-neo1). Expression was monitored by cell outgrowth in G418 for mouse P2 and P4 or by assaying the *neo*-phosphotransferase activity (*NPTII*) in protein extracts. (*P21g*$^+$, *P31g*$^+$). Lung cells from mouse P2 and P3 selected for expression in vitro; (STO$^+$) *neo*-expressing fibroblasts (positive control); (*P2br*$^-$, *P31i*$^-$) extracts from brain (P2) or liver (P3) without selection in vitro.

into the germline of mice. Since the microinjection of DNA into pronuclei does not ensure the expression of the introduced gene (Wagner et al. 1984), this system offers new possibilities for examining the action of specific gene products in mice by preselecting for the expression of these sequences in cell culture.

Gene Transfer into Bone Marrow Cells by Retrovirus Infection

Pluripotent stem cells of the hemopoietic system are another population of cells well suited to gene transfer experiments. These cells, found in adult bone marrow, can be manipulated in vitro and then transferred to hemopoietically deficient hosts where they are able to generate cells of both the myeloid and lymphoid lineages (Wu et al. 1968; Abramson et al. 1977). Although these primitive stem cells can be assayed only in long-term reconstitution experiments (Harrison 1980; Boggs et al. 1982), their descendents, the more restricted multipotent and unipotent precursors, can be detected by in vivo (Till and McCulloch 1961) and in vitro (Burgess and Nicola 1983) clonal assays.

Gene transfer into bone marrow cells using DNA-transfection has been very difficult and has become possible only recently through the use of retroviral vectors (Joyner et al. 1983; Miller et al. 1984; Williams et al. 1984).

In the first series of experiments, we tested the efficiency of infecting bone marrow cells with the *neo*-containing virus, rGag-neo (Fig. 1), using the in vitro colony assay to follow the fate of the different precursors throughout the course of the study. In a typical experiment as outlined in Table 1, approximately 15,000 colony-forming cells (CFC), were seeded onto virus-producing cells. CFC in this context refers to any cell capable of generating a colony of hemopoietic cells in a semi-solid culture under the conditions described in the legend of Table 1. Of these 15,000 CFC, 1125 were multipotent (MIX-CFC) giving rise to colonies which contain cells from the erythroid plus two other blood cell lineages. The remaining CFC were either bipotent (GM-CFC), restricted to neutrophil and macrophage differentiation, or unipotent, restricted to only neutrophil (G-CFC), macrophage (M-CFC), or mast cell (MAST-CFC) differentiation.

After viral infection of the bone marrow, 10,500 CFC were recovered, of which 1500 were G418-resistant. However, following 48 hours of culture in the presence of G418, only 3500 were recovered; but, 3000 were now resistant to G418 (Table 1). These data suggest that a high proportion of the hemopoietic cells have integrated the proviral DNA and are able to express the *neo* gene.

Virus infected cells that had been either nonselected or preselected for 48 hours in G418 were injected into lethally irradiated recipients for a short-term reconstitution experiment. The mice were killed 12 days later and the spleens were analyzed for the presence of vector DNA, expression of the *neo* protein, and assayed for neo^R CFC.

Table 1

Fate of Colony-Forming Cells Following Infection of Bone Marrow Cells with a *neo*-containing Retrovirus (rGag-neo)

Procedure	Total	Multipotent[a]	G418-resistant Total	G418-resistant Multipotent
Starting population	15,000[b]	1125	0	0
After infection	10,500	850	1500 (14%)	60 (7%)
After infection and 48-h selection in G418	3,500	175	3000 (86%)	150 (86%)

Bone marrow cells from 5-flurouracil (5-FU)-treated mice were cocultured for 24 hrs with virus-producing cells and the nonselected or preselected cells were assayed for colony-forming cells (CFC) in methyl cellulose cultures containing WEHI-3 supernatant as a source of multi HGF (Iscove 1984a,b).

[a]Multipotent refers to all precursors that give rise to colonies containing cells from the erythroid plus two other blood cell lineages.

[b]Number of CFC from two femurs of one mouse treated 5 days earlier with 150 mg/kg 5-FU.

As shown in Figure 4, vector DNA was present in the spleens of mice reconstituted with nonselected cells and to a much greater extent in the spleens of mice reconstituted with preselected cells. In addition, the introduced gene was efficiently expressed when protein extracts were assayed for the *neo*-phosphotransferase (NPTII) activity (Reiss et al. 1984). The spleen from one of these mice reconstituted with nonselected cells contained 74,000 CFC. Of these, only 4000 or 5% would grow in the presence of G418. Preselected cells used for reconstitution gave fewer CFC but a much higher proportion (70%) were neo^R. We conclude that the *neo* gene was introduced and efficiently expressed in the hemopoietic cells of the reconstituted mice.

Long term reconstitution experiments are now in progress to determine whether or not the most primitive stem cell, the cell capable of giving rise to both myeloid and lymphoid cells, have been infected with the recombinant virus.

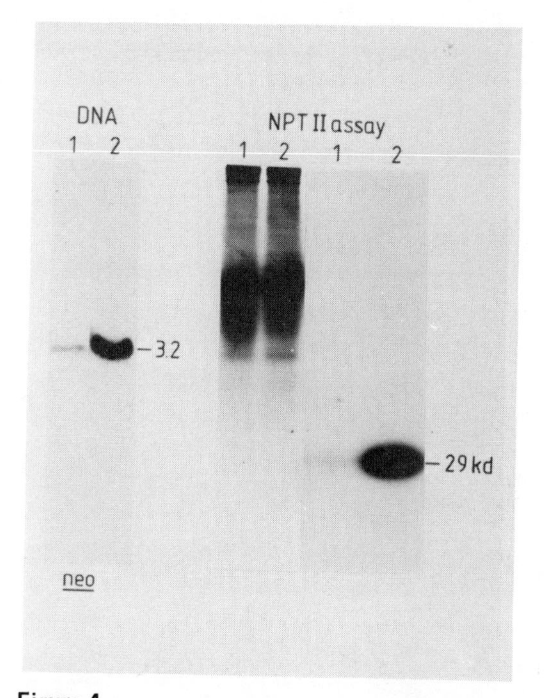

Figure 4
DNA and expression analysis of spleen cells from mice reconstituted with rGag-neo-infected bone marrow cells, which were nonselected (lane *1*) or preselected (lane *2*) in vitro. Genomic DNAs were cut with *Sac*I and hybridized with a *neo*-specific probe. The left two lanes from the NPTII assay are the proteins loaded on a nondenaturing gel before transfer onto P81 paper. The lanes on the right side visualize the presence of the active neo enzyme following an in situ hybridization assay.

Transfer and Expression of Proto-oncogenes in EC Stem Cells

In a different set of experiments we asked whether a cellular oncogene (c-*onc*) might exert a biological effect upon transfection into F9 cells. For this approach we chose the c-*fos* gene, which has previously been suggested to have a function in differentiation processes, since its expression in both fetal membranes and hemopoietic cells is correlated with differentiation (Müller et al. 1983, 1984). We found that introducing the normal mouse or human c-*fos* gene along with a selectable marker gene (*neo*) into undifferentiated F9 cells results in the appearance of cell colonies that are morphologically different from the stem cells (Müller and Wagner 1984). These cells are greatly enlarged, very flat, grow in an epitheloid fashion, and stop proliferating after reaching a maximum clone size of about 400 cells (Fig. 5B). To characterize the phenotype of the morphologically altered F9 cells, the presence of various differentiation markers was analyzed in these cells. It was found that a distinct set of markers is expressed which is different from that observed in retinoic acid-induced differentiated F9 cells (Müller and Wagner 1984). Whether the c-*fos* transfected F9 cells represent a cell type that has not yet been identified during normal development or whether an aberrant type of differentiation is triggered by c-*fos* is unknown.

To investigate the mechanism of c-*fos*-promoted differentiation in F9 cells in greater detail, recombinant vectors containing the c-*fos* coding region under the

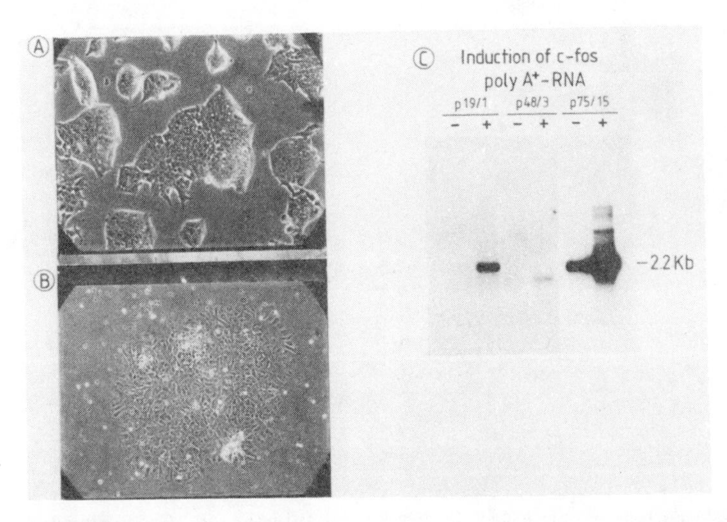

Figure 5
Phase-contrast microscopy of F9 stem cells (*A*) and morphologically altered F9 cells 10 days after transfection with an inducible MT-II-c-*fos* DNA construct (*B*). (*C*) c-*fos* RNA expression in pooled *neo*^{R} F9 clones, transfected with the different constructs (*p19/1, p48/3, p75/15*) (Rüther et al. 1985) and analyzed without (-) and with induction (+) with $CdCl_2$ for 6 hours.

control of the inducible human metallothionien promoter were constructed (Rüther et al. 1985). Following transfection into F9 cells, clonal lines carrying the different constructs were derived. Most of the lines were able to induce c-*fos* expression following a 6-hour induction period with $CdCl_2$, although constitutive expression of c-*fos* was also found (Fig. 5C). Similar results as in previous experiments were found when the extent and characteristics of differentiation was analyzed. Only a certain fraction of cells differentiated irrespective of the level of c-*fos* expression beyond a minimal threshold level. This suggests that c-*fos* alone is not sufficient and that additional, probably spontaneously occurring events are required for the induction of differentiation. Nevertheless, our findings lend support to the hypothesis that c-*fos* has a functional role in differentiation processes.

CONCLUSION

The experiments discussed in this paper have shown that gene transfer by retrovirus infection of embryonal carcinoma and hemopoietic cells offers a powerful tool to analyze gene expression both in vitro and in vivo. Our observation that the expression of exogenous c-*fos* genes can influence the differentiation of F9 teratocarcinoma stem cells suggests that future studies pertaining to the transfer of protooncogenes into stem cells and the germline of mice will provide us with novel information regarding the function of c-*onc* gene products in normal development, growth control, and in the process of neoplastic transformation.

ACKNOWLEDGMENTS

We thank Mirka Vanek for technical assistance and Ines Benner for preparing the manuscript. U.R. is a recipient of an EMBO postdoctoral long-term fellowship.

REFERENCES

Abramson, S., R.G. Miller, and R.A. Phillips. 1977. The identification in adult bone marrow of pluripotential and restricted stem cells of the myeloid and lymphoid systems. *J. Exp. Med.* **145**: 1567.

Axelrod, H.R. 1984. Embryonic stem cells lines derived from blastocysts by a simplified technique. *Dev. Biol.* **101**: 225.

Boggs, D.R., S.S. Boggs, D.F. Saxe, L.A. Gress, and D.R. Canfield. 1982. Hematopoietic stem cells with high proliferative potential. *J. Clin. Invest.* **70**: 242.

Bradley, A., M. Evans, M.H. Kaufman, and E. Robertson. 1984. Formation of germline chimaeras from embryo-derived teratocarcinoma cell lines. *Nature* **309**: 255.

Burgess, A. and N. Nicola. 1983. *Growth factors and stem cells*. Academic Press, New York.

Evans, M.J. and M.H. Kaufman. 1981. Establishment in culture of pluripotential cells from mouse embryos. *Nature* **292**: 154.

Fuji, J.T. and G.R. Martin. 1983. Developmental potential of teratocarcinoma stem cells in utero following aggregation with cleavage stage mouse embryos. *J. Embryol. Exp. Morphol.* **74**: 79.

Gautsch, J.W. and M.C. Wilson. 1983. Delayed de novo methylation in teratocarcinoma suggests additional tissue-specific mechanisms for controlling gene expression. *Nature* **301**: 32.

Graham, C.F. 1977. Teratocarcinoma cells and normal mouse embryogenesis. In *Concepts in mammalian embryogenesis* (ed. M.I. Sherman), p. 315. MIT Press, Cambridge.

Harrison, D.E. 1980. Competitive repopulation: A new assay of long-term stem cell functional capacity. *Blood* **55**: 77.

Iscove, N.N. 1984a. Specificity of hemopoietic growth factor. In *Leukemia–Dahlem Konferenzen* (ed. I.L. Weissman). Springer, Berlin.

——. 1984b. Culture of lymphocytes and hemopoietic cells in serum free medium. In *Methods for serum-free culture of neuronal and lymphoid cells,* p. 169. Alan R. Liss, New York.

Joyner, A., G. Keller, R.A. Phillips, and A. Bernstein. 1983. Retrovirus transfer of a bacterial gene into mouse hematopoietic progenitor cells. *Nature* **305**: 556.

Linney, E., B. Davis, J. Overhauser, E. Chas, and H. Fan. 1984. Non-function of a Moloney murine leukemia virus regulatory sequence in F9 embryonal carcinoma cells. *Nature* **308**: 470.

Martin, G. 1981. Isolation of a pluripotent cell line from early mouse embryos cultured in medium conditioned by teratocarcinoma stem cells. *Proc. Natl. Acad. Sci. U.S.A.* **78**: 7634.

Miller, A.D., R.J. Eckner, D.J. Jolly, T. Friedmann, and I.M. Verma. 1984. Expression of a retrovirus encoding human HPRT in mice. *Science* **225**: 630.

Mintz, B. and R.A. Fleischman. 1981. Teratocarcinomas and other neoplasms as developmental defects in gene expression. *Adv. Cancer Res.* **34**: 211.

Müller, R. and E.F. Wagner. 1984. Differentiation of F9 teratocarcinoma stem cells after transfer of c-fos proto-oncogenes. *Nature* **311**: 438.

Müller, R., I.M. Verma, and E.D. Adamson. 1983. Expression of c-onc genes: c-fos transcripts accumulate to high levels during development of mouse placenta, yolk sac and amnion. *Eur. Mol. Biol. Organ. J.* **2**: 679.

Müller, R., D. Müller, and L. Guilbert. 1984. Differential expression of c-fos in hematopoietic cells: Correlation with differentiation of monomyelocytic cells in vitro. *Eur. Mol. Biol. Organ. J.* **3**: 1887.

Mulligan, R. 1983. Construction of highly transmissable mammalian cloning vehicles derived from murine retroviruses. In *Experimental manipulation of gene expression* (ed. M. Inouye), p. 155. Academic Press, New York.

Nicolas, J.F. and P. Berg. 1983. Regulation of expression of genes transduced into embryonal carcinoma cells. In *Teratocarcinoma stem cells* (ed. L.M. Silver, G.R. Martin, and S. Strickland), Cold Spring Harbor Conf. on Cell Proliferation **10**: 469.

Niwa, O., Y. Yokota, H. Ishida, and T. Sugahara. 1983. Independent mechanisms involved in suppression of the Moloney leukemia virus genome during differentiation of murine teratocarcinoma cells. *Cell* **32**: 1105.

Reiss, B., R. Sprengel, H. Will, and H. Schaller. 1984. A new sensitive method for qualitative and quantitative assay of neomycin phosphotransferase in crude cell extracts. *Gene* **30**: 211.

Rüther, U., E.F. Wagner, and R. Müller. 1985. Analysis of the differentiation-promoting potential of inducible *c-fos* genes introduced into embryonal carcinoma cells. *Eur. Mol. Biol. Organ. J.* (in press).

Stewart, C.L. 1982. Formation of viable chimaeras by aggregation between teratocarcinoma and preimplantation mouse embryos. *J. Embryol. Exp. Morphol.* **67**: 167.

——. 1984. Teratocarcinoma chimaeras and gene expression. In *Chimaeras in developmental biology* (eds. N. Le Douarin and A. McLaren), p. 409. Academic Press, New York.

Stewart, C.L., H. Stuhlmann, D. Jähner, and R. Jaenisch. 1982. De novo methylation, expression and infectivity of retroviral genomes introduced into embryonal carcinoma cells. *Proc. Natl. Acad. Sci. U.S.A.* **79**: 4098.

Till, J.E. and E.A. McCulloch. 1961. A direct measurement of the radiation sensitivity of normal mouse bone marrow cells. *Radiat. Res.* **14**: 213.

Wagner, E.F. and B. Mintz. 1982. Transfer of nonselectable genes into mouse teratocarcinoma cells and transcription of the transferred human β-globin gene. *Mol. and Cell. Biol.* **2**: 190.

Wagner, E.F., U. Ruther, and C.L. Stewart. 1984. Introducing genes into mice and into embryonal carcinoma stem cells. In *The impact of gene transfer techniques in eucaryotic cell biology, 35th Colloquium.* Mosbach, Springer-Verlag, Berlin, Heidelberg.

Wagner, E.F., M. Vanek, and B. Vennström. 1985. Transfer of genes into embryonic carcinoma cells by retrovirus infection: Efficient expression from an internal promoter. *Eur. Mol. Biol. Organ. J.* **4**: 663.

Williams, D.A., I.R. Lemischka, D.G. Nathan, and R. Mulligan. 1984. Introduction of new genetic material into pluripotent haematopoietic stem cells of the mouse. *Nature* **310**: 476.

Wu, A.M., J.E. Till, L. Siminovitch, and E.A. McCulloch. 1968. Cytological evidence for a relationship between normal hematopoietic colony-forming cells and cells of the lymphoid system. *J. Exp. Med.* **107**: 455.

Transcription of Mouse B2 Repetitive Elements in Virally Transformed and Embryonic Cells

PETER W. J. RIGBY, PAUL M. BRICKELL, DAVID S. LATCHMAN,* NICHOLAS B. LA THANGUE, DAVID MURPHY,† AND KARL-HEINZ WESTPHAL**
Cancer Research Campaign
Eukaryotic Molecular Genetics Research Group
Department of Biochemistry
Imperial College of Science and Technology
London, SW7 2AZ, England

OVERVIEW

Mouse cells contain small RNAs which hybridize to repeated sequences of the B2 family. The abundance of these RNAs is regulated by the growth state of normal fibroblasts, by viral transformation, and by the differentiation of pluripotential embryonic cells. In vitro transcription experiments show that some B2 repeats are transcribed by RNA polymerase III and that the transforming protein of simian virus 40, large T-antigen, stimulates this transcription. We suggest that normal and embryonic cells contain proteins that are capable of regulating the transcription of B2 repeats in response to stimuli that alter growth rate or induce differentiation.

INTRODUCTION

Much of our knowledge of gene expression in eukaryotes has been derived from studies of viral systems, particularly from the intensive analyses of the DNA and RNA tumor viruses undertaken during the last two decades (Tooze 1981; Weiss et al. 1982). It is already clear that these viruses will also be useful tools with which to probe the mechanisms that regulate gene expression in the early embryo. For example, some viral promoter/enhancer regions do not function efficiently in undifferentiated embryonal carcinoma cells, but the block to expression is lifted as a result of differentiation (Kelly and Condamine 1982; Sleigh 1985). Such systems should allow the identification of interesting regulatory molecules that are likely to be involved in the control of cellular gene expression during differentiation. Several of the DNA tumor viruses, simian virus 40 (SV40), polyoma, and adenoviruses, encode, within their transforming regions, proteins that are capable of regulating

Present Addresses: *Department of Zoology, University College, London, Gower Street, London, WC1E 6BT, England; †Laboratory of Molecular Embryology, National Institute for Medical Research, The Ridgeway, Mill Hill, London, NW7 1AA, England; **Laboratorium fur Molekularbiologie, Genzentrum, Am Klopferspitz, D-8033, Martinsried, West Germany the transcription of other viral genes in *trans* (Tooze 1981). In the cases of the SV40 large T-antigen and the adenovirus Ela proteins, it is clear that they can also regulate the expression of cellular genes (Kao and Nevins 1982; Scott et al. 1983; Soprano et al. 1979). In these cases, also, it is likely that the viral proteins are mimicking cellular control mechanisms and that studies of transcriptional regulation in virus-infected and virus-transformed cells will lead us to an understanding of the normal cellular mechanisms. Our studies of the alterations in cellular transcription which occur as a result of SV40 transformation have led to the identification of a novel class of small RNAs that appear to be transcribed from B2 repetitive elements by RNA polymerase III. These RNAs also change in abundance following the differentiation of pluripotential embryonic cells. This system thus provides another example of how studies of viruses can illuminate interesting aspects of the control of cellular gene expression.

RESULTS

We have used differential cDNA cloning techniques to isolate a number of cellular genes that are expressed at higher levels in SV40-transformed mouse fibroblasts than in the parental BalB/c 3T3 cell line (Scott et al. 1983). The cDNA clones of Set 1 encode a class I antigen of the major histocompatibility complex (MHC) and contain, within the 3′ untranslated region, a member of the mouse repetitive sequence family known as B2. Because of the presence of the B2 element, these cDNA clones also hybridize to a heterogeneous collection of small RNAs that migrate on denaturing agarose gels as if they were between 700 and 400 nucleotides in length (Brickell et al. 1983). These small RNAs are present at elevated levels in cells transformed by SV40 and a number of other agents (Brickell et al. 1983; Scott et al. 1983). In order to ask whether the increased levels of these RNAs are a direct result of the expression of SV40 large T-antigen, we have studied their expression in cell lines transformed by *ts*A mutants of SV40 (Scott et al. 1983). Such mutants carry a temperature-sensitive lesion within the gene encoding large T-antigen. They will transform cells at the permissive temperature; but when such cells are shifted up to the nonpermissive temperature, the large T-antigen is inactivated and the cells revert to the normal phenotype (Tooze 1981). Figure 1 shows a Northern blot of RNA isolated from such *ts*A-transformed cells at the permissive and nonpermissive temperatures hybridized with a Set 1 cDNA probe. It is clear that the small RNAs are very strongly regulated, suggesting that their induction might be a direct effect of large T-antigen (Scott et al. 1983).

More detailed analyses have shown that the levels of these RNAs are also regulated by the growth state of normal cells. They are present at significant levels in sparse cells but decrease in abundance as the cells approach confluence (K.-H. Westphal and P.W.J. Rigby, unpubl.). Moreover, if BalB/c 3T3 cells are rendered quiescent by serum deprivation and then stimulated to reenter the growth cycle by

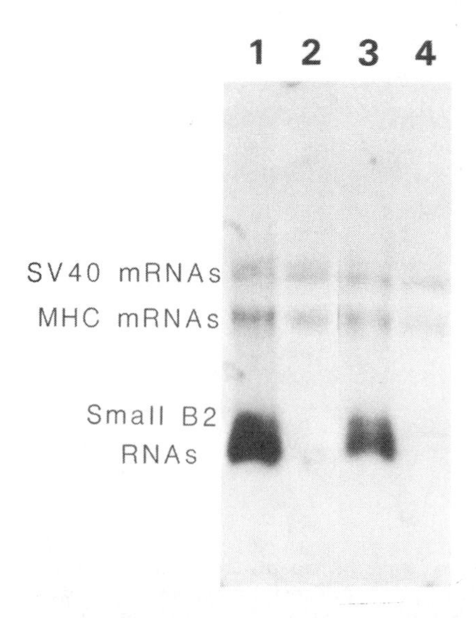

Figure 1
Transfer hybridization analysis of Set 1-related RNAs in SV40 *tsA* mutant-transformed cell lines. Polyadenylated, cytoplasmic RNA was isolated from SV3T3-A255B1b and SV3T3-A209B4a cells maintained for 4 days at either 32°C or 39.5°C. 1 μg aliquots of RNA were fractionated by electrophoresis in a 1% ($^{w}/v$) agarose gel containing formaldehyde, transferred to a nitrocellulose filter and hybridized with ^{32}P-labeled pAG64 (the prototype plasmid of Set 1) and with ^{32}P-labeled SV40 DNA. (Lane *1*) A255B1b at 32°C; (lane *2*) A255B1b at 39.5°C; (lane *3*) A209B4a at 32°C; (lane *4*) A209B4a at 39.5°C. Indicated are the SV40 mRNAs encoding large T- and small T-antigens (2.3kb and 2.6kb), the class I MHC mRNAs (1.6kb) and the small RNAs (0.7–0.4kb) detected by virtue of their homology to the B2 repetitive element. Adapted from Scott et al. (1983) in which further details are given.

feeding them with medium containing fresh serum, the level of the small RNAs increases dramatically within 15 hours (K.-H. Westphal and P.W.J. Rigby, unpubl.).

We have previously reported on the pattern of B2-containing transcripts present in pluripotential embryonic cell lines and in embryos (Murphy et al. 1983). The pattern seen in such cases is quite different to that seen in any other cell type that we have examined. Even in polyadenylated, cytoplasmic RNA, a smear of hybridization is observed, indicating the presence of a large number of transcripts containing B2 sequences. Figure 2 shows an example of this, using RNA from the pluripotential embryonic cell line EKB2B2 (Fig. 2, lane *2*). Strikingly, when we analyzed the same amount of RNA from simple embryoid bodies induced by the differentiation of EKB2B2 cells, we observed that the abundance of the polydisperse B2-containing transcripts was much reduced whereas there was a considerable increase in the level of the small RNAs (Fig. 2, lane *1*). These transcripts thus appear to be quite strongly regulated during an in vitro differentiation event, which

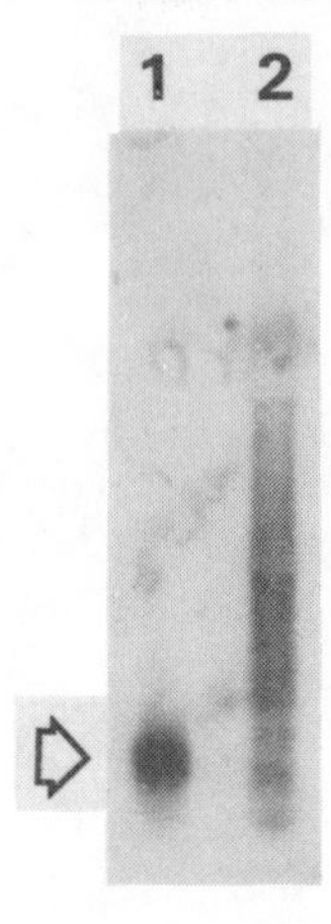

Figure 2
Transfer hybridization analysis of RNAs homologous to the B2 repetitive element in undifferentiated and differentiated EK cells. 5 μg aliquots of polyadenylated, cytoplasmic RNA were fractionated by electrophoresis in a 1% ($^{w}/v$) agarose gel containing formaldehyde, transferred to a nitrocellulose filter and hybridized with ^{32}P-labeled pAG64. (Lane *1*) EKB2B2 simple embryoid bodies; (lane *2*) undifferentiated EKB2B2 cells. The arrow indicates the small RNAs homologous to the B2 element. Adapted from Murphy et al. (1983) in which further details are given.

is thought to mimic the first overt differentiation occurring during mouse embryogenesis.

We did not recover any cDNA clones that uniquely hybridize to these small RNAs and so we do not know their precise structure. However, when Georgiev and his colleagues first characterized B2 repeats, they noticed that the element contains within it a consensus RNA polymerase III promoter (Kramerov et al. 1979; Krayev et al. 1982). We therefore wondered whether the small RNAs are polymerase III transcripts derived from B2 elements. The B2 repeat itself is approximately 190 base pairs in length, so one would have to suppose either that the small RNAs run aberrantly or that they are transcripts which originate within the repeat but extend into flanking sequences. We have asked whether cloned B2 repeats will act as promoters for RNA polymerase III transcription in vitro. We used as template the B2 element from the 3′ untranslated region of the MHC gene and also two other B2 repeats that are located upstream of the gene (P.M. Brickell, D.S. Latchman, and P.W.J. Rigby, unpubl.). Our results (N. B. La Thangue, D. Murphy, and P. W. J. Rigby, unpubl.) show that one of the upstream elements acts as an efficient promoter in a Hela cell extract and in extracts derived from a variety of mouse cell lines. Other B2 repeats are much less efficiently transcribed. If pure SV40 large T-antigen is added to the system, the amount of RNA synthesized from the upstream element increases approximately tenfold. There is certainly specificity to this stimulation as transcription from the RNA polymerase III promoter in the VA-RNA gene of adenovirus type 2 is not affected by large T-antigen.

DISCUSSION

We have thus defined a novel family of small RNAs, synthesis of which is regulated during normal growth, during the differentiation of embryonic cells, and as a consequence of transformation. Our in vitro transcription data suggest that the synthesis of these small RNAs may be directly affected by large T-antigen, and we are presently performing experiments to analyze the mechanisms involved. Presumably the regulation that occurs in normal cells proceeds by a mechanism involving cellular proteins that have at least some properties in common with large T-antigen.

These experiments provide yet another example of the validity of viral systems for studying the regulation of cellular gene expression, including events that occur during the differentiation of early embryonic cells.

ACKNOWLEDGMENTS

We are extremely grateful to Julian Gannon, Viesturs Simanis, and David Lane for providing us with pure large T-antigen. We acknowledge the support of Training Fellowships (to P.M.B. and N.B.L.) and a Research Studentship (to D.M.) from the Medical Research Council. K.-H.W. was supported during part of this work by a Long Term Fellowship from the European Molecular Biology Organisation. P.W.J.R. holds a Career Development Award from the Cancer Research Campaign which also paid for this work.

REFERENCES

Brickell, P. M., D. S. Latchman, D. Murphy, K. Willison, and P. W. J. Rigby. 1983. Activation of a *Qa/Tla* class I major histocompatibility antigen gene is a general feature of oncogenesis in the mouse. *Nature* **306**: 756.

Kao, H.-T. and J. R. Nevins. 1983. Transcriptional activation and subsequent control of the human heat shock gene during adenovirus infection. *Mol. Cell. Biol.* **3**: 2058.

Kelly, F. and H. Condamine. 1982. Tumor viruses and early mouse embryos. *Biochim. Biophys. Acta* **651**: 105.

Kramerov, D. A., A. A. Grigoryan, A. P. Ryskov, and G. P. Georgiev. 1979. Long double-stranded sequences (dsRNA-B) of nuclear pre-mRNA consist of a few highly abundant classes of sequences: Evidence from DNA cloning experiments. *Nucl. Acids Res.* **6**: 697.

Krayev, A. S., T. V. Markusheva, D. A. Kramerov, A. P. Ryskov, K. G. Skryabin, A. A. Bayev, and G. P. Georgiev. 1982. Ubiquitous transposon-like repeats B1 and B2 of the mouse genome: B2 sequencing. *Nucl. Acids Res.* **10**: 7461.

Murphy, D., P. M. Brickell, D. S. Latchman, K. Willison, and P. W. J. Rigby. 1983. Transcripts regulated during normal embryonic development and oncogenic transformation share a repetitive element. *Cell* **35**: 865.

Scott, M. R. D., K.-H. Westphal, and P. W. J. Rigby. 1983. Activation of mouse genes in transformed cells. *Cell* **34**: 557.

Sleigh, M. J. 1985. Virus expression as a probe of regulatory events in early mouse embryogenesis. *Trends Genet.* **1**: 17.

Soprano, K. J., V. G. Dev, C. M. Croce, and R. Baserga. 1979. Reactivation of silent rRNA genes by Simian virus 40 in human-mouse hybrid cells. *Proc. Natl. Acad. Sci. U.S.A.* **76**: 3885.

Tooze, J. 1981. *The molecular biology of tumor viruses. Part 2. DNA tumor viruses* (2nd ed.). Cold Spring Harbor Laboratory, Cold Spring Harbor, New York.

Weiss, R. A., N. M. Teich, H. Varmus and J. M. Coffin. 1982. *The molecular biology of tumor viruses. RNA tumor viruses* (2nd ed.). Cold Spring Harbor Laboratory, Cold Spring Harbor, New York.

Session 3: Introduction of Cloned Genes into the Mammalian Germline

Expression of Growth Hormone Genes in Transgenic Mice

RICHARD D. PALMITER,* ROBERT E. HAMMER,† AND
RALPH L. BRINSTER†
*Howard Hughes Medical Institute
University of Washington
Seattle, Washington 98195
†School of Veterinary Medicine
University of Pennsylvania
Philadelphia, Pennsylvania 19104

OVERVIEW

Human or rat growth hormone (GH) genes have been introduced into all cells of a mouse by microinjection of fertilized eggs but they were not expressed under their own promoters. However, substitution of a mouse metallothionein (MT) promoter allowed expression and regulation comparable to that of the endogenous MT genes. These fusion genes have been used to stimulate the growth of both normal mice and dwarf mice that lack sufficient GH. Substitution of a rat elastase-I promoter directed expression of GH exclusively to the acinar cells of the pancreas. Progress has been made towards developing the hGH gene into a vector that is not expressed in vivo unless an enhancer element is inserted. Recombination between overlapping DNA fragments derived from a MThGH gene, each of which is nonfunctional, has been observed when they are coinjected into mouse eggs. In some cases, functional hGH was produced as evidenced by enhanced growth of the mice.

INTRODUCTION

Growth hormone (GH) is an intermediary in a cascade of hormones that control growth of mammals. It is a single chain polypeptide of 191 amino acids that is synthesized by sommatotroph cells in the pituitary. Hypothalamic hormones, somatostatin, and growth hormone releasing factor, regulate GH synthesis; GH, in turn, regulates the production of insulin-like growth factor-I (IGF-I) by peripheral tissues (Palmiter et al. 1983). GH genes have been isolated from several species (Barta et al. 1981; Seeburg 1982; Gordon et al. 1983). They are composed of five exons and span a total of 2-3 kb; thus, they are of a convenient size for genetic manipulation. They are related to placental lactogen and prolactin genes (Niall et al. 1971).

Several groups have introduced genes into the germline of mice by microinjecting appropriate DNA fragments isolated from plasmids into the pronuclei of fertilized eggs (Gordon et al. 1980; Brinster et al. 1981; Costantini and Lacy 1981; E. Wagner et al. 1981; T. Wagner et al. 1981). With current techniques, about 25% of the mice that develop from this procedure retain the foreign DNA in all of their cells and transmit them to half of their offspring. Here, we summarize the results obtained with GH genes and indicate some of the future directions.

RESULTS AND DISCUSSION

Metallothionein-growth Hormone Fusion Genes Stimulate Growth of Mice

Because of our initial success in obtaining regulated expression of thymidine kinase gene by fusing the mouse metallothionein-I (MT) promoter to the structural gene isolated from herpes simplex virus (Brinster et al. 1981; Palmiter et al. 1982a), we initiated our experiments with GH in a similar manner. In the first experiments, the MT promoter was fused to the structural gene of rat GH (rGH), and DNA fragments retaining 185 bp of MT promoter and all of the rGH structural gene were microinjected into the pronuclei of fertilized eggs. Of the 21 mice that developed from these eggs, six grew significantly larger than control littermates, and several of these mice had extraordinarily high levels of GH mRNA in the liver and rGH in the serum (Palmiter et al. 1982b). Several lines of mice were started from these transgenic founders. One of these lines, MGH-10, is now in the sixth generation; about 50% of the offspring inherit the chromosome carrying the MTrGH genes and all of these mice grow to about twice the size of normal littermates (Table 1).

These mice grow because the mouse MT promoter causes GH to be synthesized in several organs, notably liver and kidney, instead of in the sommatotroph cells of the pituitary. Although the cellular rate of GH production in these transgenic mice may be lower than in the sommatotroph cells, the enormous size of these organs compared to the pituitary allows serum concentrations to reach levels that are 1000-fold higher than normal (Palmiter et al. 1982b). The production of rGH can be modulated about tenfold by adding zinc, a natural inducer of MT genes, to the diet (Hammer et al. 1984a). However, this extra stimulation of GH synthesis is not required to stimulate growth, presumably because the basal rate of synthesis is sufficient to saturate GH receptors.

Table 1
Effects of MTrGH Gene Expression on Growth and Fertility of Normal (C57 X SJL Hybrids) and Little (lit/lit) Mice[a]

	Males		Females	
Mice	Adult size	Fertility	Adult size	Fertility
Little	15 g	+/-	13 g	++
Little · MTrGH	43 g	++	41 g	+/-
Normal	27 g	+++	23 g	+++
Normal · MTrGH	47 g	++	39 g	+/-

[a]See Palmiter et al. (1982a) and Hammer et al. (1984a,b) for details.

Partial Correction of a Genetic Disease

One of the original attractions of GH genes for expression in transgenic mice was the existence of several mutant strains of mice with defects in GH production. One dwarf strain, called little, grows to about half normal size when homozygous (lit/lit). Although the primary defect that leads to suboptimal GH production in little mice is unknown, injection of GH will stimulate growth (Beamer and Eicher 1976). Therefore, we reasoned that the phenotypic defects in growth could probably be overcome by introducing MTrGH fusion genes into fertilized (lit/lit) eggs. Table 1 shows that enhanced growth was achieved by using this approach (Hammer et al. 1984b).

Homozygous (lit/lit) males show a high degree of infertility, whereas females eventually reach full fertility. The fertility of male little mice expressing MTrGH genes was also corrected; all transgenic males have sired at least two litters. However, the fertility of females (either lit/lit or wild-type) that express MTrGH genes was impaired (Table 1 and Hammer et al. 1984b).

Expression of Human Growth Hormone and Placental Lactogen Genes

In humans, there is a cluster of five GH-related genes located within 45 kb of DNA on chromosome 17. These genes have been isolated on two cosmids (Barsh et al. 1983); one of these, cGH4, contains the normal hGH gene, a placental lactogen-like gene (hPL_L) and a normal placental lactogen gene (hPL_A). None of sixteen transgenic mice carrying this cosmid showed enhanced growth, serum hPL, or hGH as measured by RIA (Fig. 1). Another 16 transgenic mice with plasmids containing either hGH or rGH genes also failed to express these genes (Hammer et al. 1984b). Thus, it appears that the signals necessary for proper expression of human or rat GH genes are either absent from the DNA molecules tested thus far or they are incapable of responding to mouse regulatory factors when introduced into the germline by microinjection.

We have achieved expression of hGH_N and hPL_A by fusing these structural genes to the mouse MT promoter in a manner similar to that used for rGH (Fig. 1). MThGH fusion genes work as well as MTrGH genes at stimulating the growth of mice (Palmiter et al. 1983). The foreign gene is expressed predominantly in liver, heart, testis, and intestine; but measurable levels of MThGH mRNA are detectable in other tissues as well. The pattern of expression resembles that of the endogenous MT genes (Palmiter et al. 1983).

One line of mice expressing a high level (about 7 μg/ml in the serum) of hPL_A has been examined in detail to see if this hormone has any effect upon murine growth or reproductive physiology. The fertility of males and females expressing this gene is normal; fetal and adolescent growth are normal; and maternal behavior and lactation are normal. Thus, we have been unable to help unravel the mystery concerning the physiological role of hPL (Chard 1983).

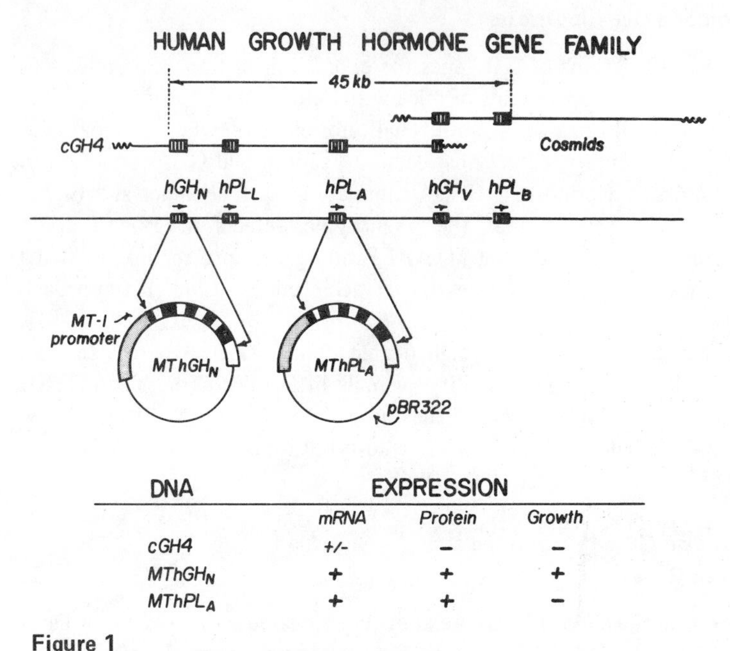

DNA	EXPRESSION		
	mRNA	Protein	Growth
cGH4	+/-	−	−
$MThGH_N$	+	+	+
$MThPL_A$	+	+	−

Figure 1
Organization of the human GH gene family and expression of hGH_N and hPL_A in transgenic mice. Cosmid, cGH4, and two hybrid genes with the mouse MT-I gene promoter fused to the structural gene of hGH_N or hPL_A were microinjected into mouse eggs. Expression of these genes in transgenic mice was tested by monitoring growth, hGH and hPL mRNA levels, and by radioimmune assay of hGH and hPL in serum samples. Very low levels of hGH mRNA were detected in liver and pituitary samples from a few mice carrying the cGH4 cosmid DNA, but no serum proteins corresponding to hGH or hPL were detectable. See Palmiter et al. (1983) for details regarding $MThGH_N$ gene expression.

Development of the hGH Gene into an Enhancer Vector

To help define the DNA sequences involved in cell-specific gene expression, it would be useful to have a gene that is not expressed at all in vivo unless appropriate sequences, so-called enhancer elements, are supplied. If the vector coded for a gene product that had a pronounced physiological effect when expressed in any cell type, then one could test various enhancers in a common vector. Expression could initially be monitored by the physiological effect and then the source of the gene product could be tracked to the cell type of origin. The data presented above suggested that GH genes might be adaptable to this purpose since they were not expressed under their own promoters but produced a readily apparent physiological effect when expressed under control of a heterologous promoter. An advantage of GH over many other secreted hormones is that the only post-translational modification involves removal of the signal peptide. Thus, it should be possible to produce and secrete functional GH in any cell type with signal peptidase.

Our approach was to introduce DNA fragments containing enhancer function upstream of the hGH promoter. We chose the mouse MT-1 gene region located between -46 and -185, which includes the entire cluster of metal regulatory elements (MREs), as a test enhancer (Stuart et al. 1984). The various constructs were tested in tissue culture cells by standard transfection and transient assay or by microinjecting them into fertilized eggs and testing for expression in the liver of fetuses or adults. Figure 2 shows that the tissue culture assay was somewhat permissive and allowed a low level of expression (measured as hGH mRNA) whereas the in vivo test system was more stringent and no expression could be detected in the absence of the test enhancer. Insertion of the MRE enhancer region (Fig. 2, solid box) about 300 bp upstream of the hGH cap site had no effect. Likewise, several other enhancers, including the SV40 72-bp repeats, had no effect in vivo when inserted at -300 (data not shown). However, when the MRE region was moved to within 90 bp of the hGH cap site, then expression in tissue culture cells could be detected and hGH mRNA was inducible by metals. However, this construct was still ineffective in vivo unless the vector sequences were removed. The control (hGH_{-90}) without vector sequences is inactive in vivo. We are in the process of testing other enhancers to see if this construct has general utility as an enhancer vector. It appears that DNA sequences lying between -90 and -300 of the hGH gene prevent MRE enhancer activation of hGH transcription and that this gene is very sensitive to vector (pUC) sequences in vivo.

DNA	EXPRESSION	
	tissue culture	*mice*
MThGH$_{+2}$ (*hGH*)	+++	+++
hGH$_{-300}$	+	-
(M) hGH$_{-300}$	+	-
(M)hGH$_{-90}$	+++	-
(M) hGH$_{-90}^{-}$ V	*ND*	++

Figure 2
Development of an hGH enhancer vector. MThGH is a hybrid gene with the mouse MT-I promoter fused to the hGH structural gene (arrow) at the BamHI site (+2) as shown in Figure 1; the solid box represents the metal regulatory elements (MRE; -46 to -185) of the mouse MT-I gene; and the wavy line represents pUC vector sequences. hGH_{-300} consists of the hGH gene with about 300 bp of 5′ flanking sequence. (M)hGH_{-300} has the MRE inserted at -300. (M)hGH_{-90} has the MRE inserted at -90. The last construction lacks vector (V) sequences. To monitor expression, DNA was either transfected into baby hamster kidney cells and hGH mRNA was measured 48 hr later, or DNA was microinjected into mouse eggs and hGH mRNA was assayed in fetal or adult liver.

Use of hGH to Study Cell-specific Gene Expression

The rat elastase-I gene is expressed in the acinar cells of the pancreas. When this gene was introduced into the germline of mice, rat elastase was still expressed almost exclusively in the pancreas (Swift et al. 1984). Deletion of all but 205 bp of 5′ flanking sequences still allowed pancreas-specific expression but the absolute level of expression was somewhat lower (Fig. 3). To determine whether 5′ rat elastase sequences were sufficient to direct expression to the pancreas, we inserted a convenient linker at +8 of rat elastase gene and fused the 5′ flanking sequences to the +2 position of hGH. This fusion gene was also expressed in a pancreas-specific manner when either 4500 bp or 205 bp of rat elastase sequences were present (Fig. 3). The level of expression of either rat elastase or hGH was at least three orders of magnitude higher in pancreas than in any other tissue tested and in many cases it was five orders of magnitude higher (Swift et al. 1984; Ornitz et al. 1985). Furthermore, the level of foreign mRNA produced frequently exceeded 10^4 molecules/cell. We do not yet know whether the enhancer and promoter functions are separable, but if so then it appears that the elastase promoter is very tightly controlled by its associated enhancer.

It is interesting to note that the transgenic mice expressing the elastase-hGH fusion genes did not show any signs of enhanced growth despite the high level of expression in the pancreas. Immunofluorescent analysis of hGH in sections of the pancreas from these animals revealed intense fluorescence over the acinar cells and in the collecting ducts; but islets, lymph nodes, and capillaries were negative (Ornitz et al. 1985). Thus, we suspect that hGH was secreted along with the digestive enzymes into the gut and none was resorbed intact or secreted into the circulation. In fact, the absence of a growth effect argues strongly that these mice were not synthesizing hGH in any tissue that secretes into the bloodstream. This is one limitation of hGH as the ideal enhancer vector as discussed in the previous section.

DNA	EXPRESSION pancreas	EXPRESSION other tissues
elastase (-7kb) — elastase	+++	0
elastase (-0.2) — elastase	++	0
elastase (-4.5) — hGH	+++	0
elastase (-0.2) — hGH	++	0

Figure 3
Tissue-specific expression of rat elastase and elastase-hGH fusion genes. The rat elastase-I gene with 7.0 kb or 0.2 kb of 5′ flanking sequence was expressed almost exclusively in the pancreas (Swift et al. 1984). Hybrid genes with 4.5 kb and 0.2 kb of elastase 5′ flanking sequences fused to the hGH structural gene were also expressed exclusively in the pancreas (Ornitz et al. 1985).

Use of MThGH to Study Recombination

One of our goals is to obtain some control over the integration of foreign DNA. At present we have little control over the number of copies that integrate or the site at which they integrate. The integration frequency is improved about fivefold by injecting linear molecules compared to circular forms (Brinster et al. 1985). There is usually a single integration site and if there is more than one copy of the foreign DNA integrated they are usually in a tandem head-to-tail array. The tandem arrays could result from homologous recombination between the injected molecules either before or during the integration process (Brinster et al. 1981). However, we have not observed homologous recombination between the injected DNA and homologous endogenous genes (R. Palmiter et al., unpubl.).

We constructed the MThGH vectors shown in Figure 4 as a means of studying recombination. First, we inserted a FLP sequence from the 2-micron circle of yeast into the third intron of MThGH (Fig. 4, solid circle) because an enzyme (FLPase) that will promote site-specific recombination at this sequence can be isolated (Cox 1983; Meyer-Leon et al. 1984). This enzyme may ultimately be useful for targeting foreign DNA to specific sites. Insertion of the FLP sequence had no effect on MThGH expression (Fig. 4). Then we deleted several hundred base pairs between *Nar*I and *Sma*I (plasmid #165), which should result in a truncated hGH protein of 104 amino acids. We have not yet looked for this protein, but we know that mRNA levels are high and that the mice do not grow larger than normal; we suspect that

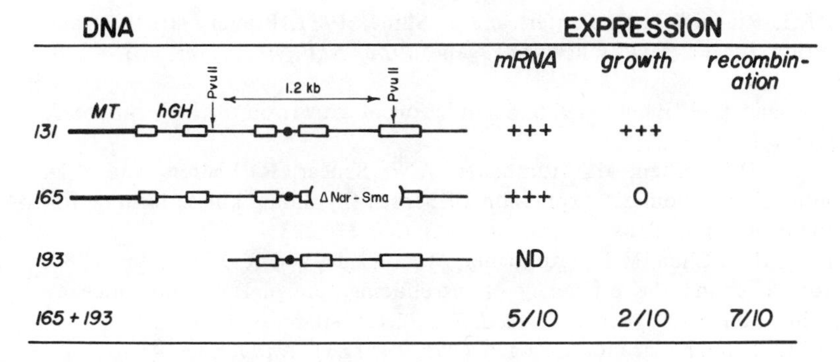

Figure 4
MThGH constructions designed to test for recombination. Plasmid #131 (top line) is a MThGH fusion gene as shown in Figure 1 with the addition of a FLP sequence (•) in the third intron. Plasmid #165 has a deletion between the *Nar*I and *Sma*I sites which deletes the last 87 amino acids from hGH and the 3′ *Pvu*II site. Fragment #193 lacks the MT promoter, the first two exons of hGH, and the 5′ *Pvu*II site. Coinjection of a few hundred copies of #165 and #193 into fertilized eggs resulted in ten transgenic mice: seven of them showed evidence of the correct 1.2-kb *Pvu*II fragment; five expressed RNA sequences present only in #193; and two of them grew significantly larger than controls.

this protein is made but is biologically inactive. The ultimate experiment will be to supply the correct information to replace the deleted nucleotides by introducing DNA fragment #193 along with FLPase into eggs from mice carrying resident copies of fragment #165. As a control for this experiment, we coinjected fragments #165 and #193. Ten mice resulting from this experiment retained these plasmid sequences. Southern blots of their DNA revealed the presence of a 1.2 kb *Pvu*II fragment that is indicative of recombination between the two DNA molecules. Five of these mice had mRNA sequences corresponding to the region deleted from fragment #165, and two of them grew significantly larger than their control littermates. Recombination between DNA molecules that are introduced into cells simultaneously has been reported (Folger et al. 1982; DeSaint Vincent and Wahl 1983; Shapira et al. 1983; Small and Scangos 1983; Subramani and Berg 1983), but in most of those experiments there was selection for the recombination event. In this experiment there was no selection, yet the frequency of recombination was about 70%. Furthermore, in two of these cases the recombination occurred in a manner that allowed production of functional hGH. In the other three mice that showed evidence of gene expression but failed to grow larger than normal, the recombination events may have been imprecise and thereby resulted in an aberrant hGH protein.

REFERENCES

Barsh, G.S., P.H. Seeburg, and R.E. Gelinas. 1983. The human growth hormone gene family: Structure and evolution of the chromosomal locus. *Nucl. Acids Res.* **11**: 3939.

Barta, A., R.I. Richards, J.D. Baxter, and J. Shine. 1981. Primary structure and evolution of rat growth hormone gene. *Proc. Natl. Acad. Sci. U.S.A.* **78**: 4867.

Beamer, W.E. and E.M. Eicher. 1976. Stimulation of growth in the little mouse. *J. Endocrinol.* **71**: 37.

Brinster, R.L., H.Y. Chen, M. Trumbauer, A.W. Senear, R. Warren, and R.D. Palmiter. 1981. Somatic expression of herpes thymidine kinase in mice following injection of a fusion gene into eggs. *Cell* **27**: 223.

Brinster, R.L., H.Y. Chen, M.E. Trumbauer, M.K. Yagle, and R.D. Palmiter. 1985. Factors affecting the efficiency of introducing foreign DNA into mice by microinjecting eggs. *Proc. Natl. Acad. Sci. U.S.A.* (in press).

Chard, T. 1983. Human placental lactogen. *Curr. Top. Exp. Endocrinol.* **4**: 167.

Costantini, F. and E. Lacy. 1981. Introduction of rabbit β-globin gene into the mouse germ line. *Nature* **294**: 92.

Cox, M.M. 1983. The FLP protein of yeast 2μm plasmid: Expression of a eukaryotic genetic recombination system in *Escherichia coli*. *Proc. Natl. Acad. Sci. U.S.A.* **80**: 4223.

DeSaint Vincent, B.R. and G. Wahl. 1983. Homologous recombination in mammalian cells mediates formation of a functional gene from two overlapping gene fragments. *Proc. Natl. Acad. Sci. U.S.A.* **80**: 2002.

Folger, K.R., E.A. Wong, G. Wahl, and M.R. Capecchi. 1982. Pattern of integration of DNA microinjected into cultured mammalian cells: Evidence for homologous recombination between injected plasmid DNA molecules. *Mol. Cell. Biol.* **2**: 1372.

Gordon, J.W., G.A. Scangos, D.J. Plotkin, J.A. Barbosa, and F.H. Ruddle. 1980. Genetic transformation of mouse embryos by microinjection of purified DNA. *Proc. Natl. Acad. Sci. U.S.A.* **77**: 7380.

Gordon, D.F., D.P. Quick, C.R. Erwin, J.E. Donelson, and R.A. Maurer. 1983. Nucleotide sequence of the bovine growth hormone chromosomal gene. *Mol. Cell. Endocrinol.* **33**: 81.

Hammer, R.E., R.D. Palmiter, and R.L. Brinster. 1984a. Introduction of metallothionein-growth hormone fusion genes into mice. In *Advances in gene technology: Human genetic disorders* (ed. F. Ahmad et al.), vol. 1, p. 52. ICSU Press, Miami, Florida.

———. 1984b. Partial correction of murine hereditary disorder by germ-line incorporation of a new gene. *Nature* **311**: 61.

Meyer-Leon, L., J.F. Senecoff, R.C. Bruckner, and M.M. Cox. 1984. Site-specific genetic recombination promoted by the FLP protein of yeast 2-micron plasmid in vitro. *Cold Spring Harbor Symp. Quant. Biol.* **49**: 797.

Niall, H.D., M. Hogan, K. Sauer, I. Rosenblum, and F. Greenwood. 1971. Sequences of pituitary and placental lactogenic growth hormones. Evolution from a primordial peptide by gene reduplication. *Proc. Natl. Acad. Sci. U.S.A.* **68**: 866.

Ornitz, D.M., R.D. Palmiter, R.E. Hammer, R.L. Brinster, G.H. Swift, and R.J. MacDonald. 1985. Specific expression of an elastase-human growth hormone fusion gene in pancreatic acinar cells of transgenic mice. *Nature* **313**: 600.

Palmiter, R.D., H.Y. Chen, and R.L. Brinster. 1982a. Differential regulation of metallothionein-thymidine kinase fusion genes in transgenic mice and their offspring. *Cell* **29**: 701.

Palmiter, R.D., R.L. Brinster, R.E. Hammer, M.E. Trumbauer, M.G. Rosenfeld, N.C. Birnberg, and R.M. Evans. 1982b. Dramatic growth of mice that develop from eggs microinjected with metallothionein-growth hormone fusion genes. *Nature* **300**: 611.

Palmiter, R.D., G. Norstedt, R.E. Gelinas, R.E. Hammer, and R.L. Brinster. 1983. Metallothionein-human GH fusion genes stimulate growth of mice. *Science* **222**: 809.

Seeburg, P.H. 1982. The human growth hormone gene family: Nucleotide sequences show recent divergence and predict a new polypeptide hormone. *DNA* **1**: 239.

Shapira, G., J.L. Stachelek, A. Letsou, L.K. Soodak, and R.M. Liskay. 1983. Novel use of synthetic oligonucleotide insertion mutations for study of homologous recombination in mammalian cells. *Proc. Natl. Acad. Sci. U.S.A.* **80**: 4827.

Small, J. and G. Scangos. 1983. Recombination during gene transfer into mouse cells can restore the function of deleted genes. *Science* **219**: 174.

Stuart, G.W., P.F. Searle, H.Y. Chen, R.L. Brinster, and R.D. Palmiter. 1984. A twelve base pair DNA motif that is repeated several times in metallothionein gene promoters confers metal regulation to a heterologous gene. *Proc. Natl. Acad. Sci. U.S.A.* **81**: 7318.

Subramani, S. and P. Berg. 1983. Homologous and nonhomologous recombination in monkey cells. *Mol. Cell. Biol.* **3**: 1040.

Swift, G.H., R.E. Hammer, R.J. MacDonald, and R.L. Brinster. 1984. Tissue-specific expression of the rat pancreatic elastase I gene in transgenic mice. *Cell* **38**: 639.

Wagner, E., T. Stewart, and B. Mintz. 1981. The human β-globin gene and a functional viral thymidine kinase gene in developing mice. *Proc. Natl. Acad. Sci. U.S.A.* **78**: 5016.

Wagner, T.E., P.C. Hoppe, J.D. Jollick, D.R. Scholl, R. Hodinka, and J.G. Gault. 1981. Microinjection of rabbit β-globin gene into zygotes and its subsequent expression in adult mice and their offspring. *Proc. Natl. Acad. Sci. U.S.A.* **78**: 6376.

Production of Tumors Following the Introduction of T-Antigen Genes into Mice

RALPH L. BRINSTER * AND RICHARD D. PALMITER†
*School of Veterinary Medicine
University of Pennsylvania
Philadelphia, Pennsylvania 19104
†Howard Hughes Medical Institute
University of Pennsylvania
Seattle, Washington 98195

INTRODUCTION

A variety of foreign gene constructions have been introduced into mice by microinjecting DNA into the pronucleus of the fertilized egg (see Palmiter et al., this volume). The resulting animals are called transgenic and frequently express the new gene. Our success in obtaining expression of the metallothionein-thymidine kinase (MK) and metallothionein-growth.hormone (MGH) genes in mice (Brinster et al. 1981; Palmiter et al. 1982, 1983), led us to examine the effect of introducing transforming or *onc* genes into mice.

DISCUSSION

An interesting and well-characterized set of transforming genes is present in the early region of the simian virus 40 (SV40) genome which contains both the large T-antigen and small T-antigen coding sequences. In earlier studies (Jaenisch and Mintz 1974), the injection of SV40 virus or DNA into the blastocoel cavity of mice resulted in animals containing the SV40 genome, but no tumors were found in these animals. In an attempt to increase expression of the SV40 transforming genes after introduction into transgenic mice, we inserted the SV40 early region into a plasmid containing a metallothionein fusion gene known to be expressed in transgenic mice, and injected this construction into eggs. The resulting mice developed characteristic choroid plexus papillomas and carcinomas (Brinster et al. 1984).

To gain insight into which DNA elements may be responsible for tissue-specific expression and tumorigenic potential, we have modified the construction of the injected DNA. Four separate modifications have been tested. First, the companion gene (MK or MGH) was removed, leaving only the T-antigen genes with their intact regulatory region. Mice containing this construction develop essentially the same pathology as previously described. Thus, the companion gene is not necessary for development of choroid plexus tumors. Second, the 72 bp enhancers were removed from the original construction (SV-MGH). Mice containing the gene without enhancer rarely develop choroid plexus tumors but develop a peripheral neuropathy with limb paralysis, liver tumors, and pancreatic adenomas. Therefore, the enhancer appears to play a significant role in targeting the pathology to the choroid plexus. Third, deletions were made in the early region to remove essential parts of either the large T- or small T-antigen. These studies indicated that choroid plexus tumors developed when only the large T-antigen gene, but not when only the small T-antigen gene, was introduced into mice. Fourth, the rat elastase I gene 5′ flanking region (Swift et al. 1984; Palmiter et al., this volume) was fused to the SV40 structural gene (lacking 72 bp enhancers and 21 bp repeats), and this construction was introduced into mice. Several of these animals developed pancreatic adenocarcinomas with little sign of other pathology. Thus, the elastase promoter/enhancer appears to be able to target the cancer producing ability of the SV40 with great specificity.

CONCLUSION

In summary, a transgenic mouse model system has been developed that will allow an identification of the DNA sequences in the SV40 early region necessary for in vivo cell transformation and cancer development. Furthermore, the system can be exploited to study mechanisms involved in targeting cancer formation to specific tissues.

REFERENCES

Brinster, R.L., H.Y. Chen, M.E. Trumbauer, A.W. Senear, R. Warren, and R.D. Palmiter. 1981. Somatic expression of herpes thymidine kinase in mice following injection of a fusion gene into eggs. *Cell* **27**: 223.

Brinster, R.L., H.Y. Chen, A. Messing, T. vanDyke, A.J. Levine, and R.D. Palmiter. 1984. Transgenic mice harboring SV40 T-antigen genes develop characteristic brain tumors. *Cell* **37**: 367.

Jaenisch, R. and B. Mintz. 1974. Simian virus 40 DNA sequences in DNA of healthy adult mice derived from preimplantation blastocysts injected with viral DNA. *Proc. Natl. Acad. Sci. U.S.A.* **71**: 1250.

Palmiter, R.D., R.L. Brinster, R.E. Hammer, M.E. Trumbauer, M.G. Rosenfeld, N.C. Birnberg, and R.M. Evans. 1982. Dramatic growth of mice that develop from eggs microinjected with metallothionein-growth hormone fusion genes. *Nature* **300**: 611.

Palmiter, R.D., G. Norstedt, R.E. Gelinas, R.E. Hammer, and R.L. Brinster. 1983. Metallothionein-human growth hormone fusion genes stimulate growth of mice. *Science* **222**: 809.

Swift, G.H., R.E. Hammer, R.J. MacDonald, and R.L. Brinster. 1984. Tissue-specific expression of the rat pancreatic elastase I gene in transgenic mice. *Cell* **38**: 639.

Heritable Oncogenesis in Transgenic Mice Harboring a Recombinant Insulin/SV40 T-Antigen Gene

DOUGLAS HANAHAN
Cold Spring Harbor Laboratory
Cold Spring Harbor, New York 11724

OVERVIEW

A recombinant oncogene composed of DNA derived from the 5′ region of the rat insulin (II) gene linked to protein coding information for SV40 T-antigen has been transferred into the mouse germline. T-antigen is found exclusively in the insulin-producing β cells of the endocrine pancreas. Expression of this recombinant oncogene elicits hyperplasia of the islets of Langerhans, and the subsequent heritable development of solid β cell tumors.

INTRODUCTION

Techniques for establishing cloned genes in the mouse germline have reached fruition during the last several years. DNA injected into fertilized one cell mouse embryos becomes integrated into a chromosome with appreciable frequency–about 10% of mice born from such embryos carry the injected DNA of a heritable genetic element. Mice harboring transferred genes have been used to assess the ability of these genes to express correctly as a consequence of the *cis*-acting elements associated with them. Such analyses have shown that an increasing number of genes can express preferentially in the correct tissue type following "random" integration into the mouse germline. These include genes (or promoter elements derived from genes) encoding metallothionein (Palmiter et al. 1982a), immunoglobulins (Brinster et al. 1983; Grosschedl et al. 1984), transferrin (McKnight et al. 1983), elastase (Swift et al. 1984), and β-globin (Chada et al. 1985).

A second application of DNA transfer into mice has involved the examination of the consequences of the expression of the newly acquired genes. When the structural information for rat growth hormone was linked to the promoter element for metallothionein and transferred into mice, the fusion gene preferentially expressed in liver and kidney and the synthesis of rat growth hormone mediated the development of unusually large mice (Palmiter et al. 1982b). The SV40 early region elicits the formation of choroid plexus tumors in transgenic mice, and oncogenesis appears to involve a gene activation event, which effects high-level expression of a T-antigen gene that appears to be inactive in normal tissue (Brinster et al. 1984). The proto oncogene *c-myc*, when fused to the long terminal repeat of mouse mammary tumor virus, produces breast tumors in transgenic mice, in a manner that suggests that secondary events are necessary to elicit tumor formation (Stewart et al. 1984). More recently, the SV40 T-antigen gene was linked to the 5′ flanking region of the rat elastase gene, and transgenic mice harboring this hybrid gene were found to develop pancreatic adenomas (Palmiter et al., this volume). Taken together, these studies indicate that gene transfer into mice by microinjection of fertilized one cell embryos can be used to study the control of gene expression and the consequences of such expression.

The general approach taken here is intended to address both the requirements for correct tissue-specific expression of developmentally regulated genes, and the consequences of tissue-specific expression of oncogenes. The strategy employed is to construct recombinant genes composed of putative regulatory information associated with tissue-specific genes linked to protein coding information for oncogenes, and to establish these hybrid genes in the mouse germline so as to examine the characteristics and consequences of their expression.

The recombinant oncogene described in this paper utilizes the 5′ flanking region of the rat insulin (II) gene and the protein coding information of the SV40 early region, which encodes large T- and small T-antigens. The insulin genes are expressed only in the β cells of the endocrine pancreas, and thus represent genes with a very tight specificity, being active only in one cell type of one tissue (Steiner and Freinkel 1972; Cooperstein and Watkins 1981). The rat insulin genes have been cloned and sequenced (Ullrich et al., 1977; Villa-Komaroff et al. 1978; Cordell et al. 1979; Lomedico et al. 1979). The 5′ flanking region of the rat insulin I gene has been fused to a bacterial gene and shown to mediate transient expression in cultured insulinoma cells, but not in fibroblasts (Walker et al. 1983). This indicates that hybrid genes employing the insulin gene 5′ flanking region can mediate expression in an appropriate (insulin producing) cultured cell, and that the body of the insulin gene is not obligatory for expression, in contrast to observations that β-globin genes (Charnay et al. 1984; Wright et al. 1984) and immunoglobulin genes (R. Grosschedl and D. Baltimore, in prep.) are not readily adaptable to hybrid genes that substitute different coding regions. SV40 T-antigen is a 96kd nuclear protein which is capable of both immortalizing primary cells to growth in culture and of releasing normal cells from dependence on serum growth factors and contact inhibition, and thereby promoting their uncontrolled proliferation (Tooze 1981). T-antigen is also a good antigen that is not normally present in mouse cells, and thus immunological reagents can be used to assess its presence in the tissues of transgenic mice.

RESULTS

Hybrid Gene Structure and Generation of Transgenic Mice

A hybrid gene was constructed so as to combine sequences 5′ to the rat insulin (II) gene with protein coding information for SV40 large T-antigen. The structural information for large T-antigen was contained in a fragment derived from the early region of SV40 that begins in the 5′ untranslated region of early transcription and includes protein coding information for large T and small T proteins, which are derived by differential splicing, as well as sequences that mediate transcriptional termination and polyadenylation of nascent mRNAs. This fragment lacks the SV40 early promoter and enhancer region, and thus does not comprise a complete transcription unit.

DNA 5′ to the rat insulin (II) gene was fused to the T-antigen structural information so as to provide elements for initiation of transcription. The insulin 5′ flanking region includes a promoter, which mediates initiation of transcription of the insulin gene, as well as a transcriptional enhancer element which stimulates initiation of transcription in cultured insulinoma cells but not in fibroblasts (Walker et al. 1983). The insulin control region was aligned so as to promote transcription of the T-antigen gene, as diagrammed in Figure 1. The two genes were recombined in regions corresponding to the 5′ untranslated portion of each mRNA. The resulting hybrid gene (RIP1-Tag) thus comprises a complete transcription unit with 660 bp of sequence located 5′ to the point of transcriptional initiation of the insulin gene linked to protein coding and termination information from the early region of SV40.

The fusion gene was injected into fertilized one cell mouse embryos (Gordon et al. 1980; Palmiter 1982a; Lacy et al. 1983), which were then inserted into the oviducts of pseudopregnant female mice and allowed to develop. Four transgenic mice were produced from these injections. Two of the founder mice (165 and 168) were

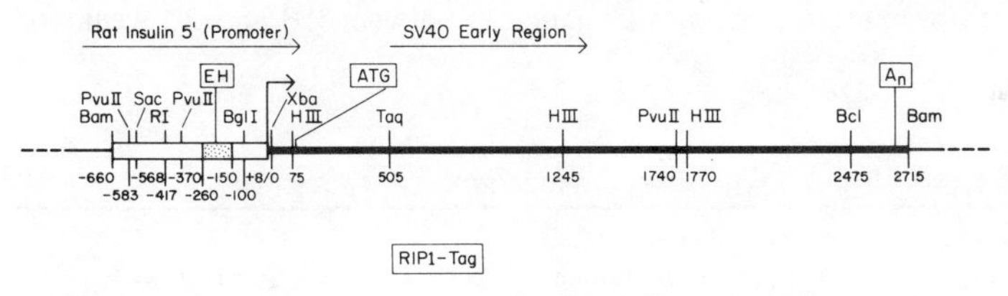

Figure 1
Structure of recombinant insulin/T-antigen genes. The plasmid pRIP1-Tag consists of the coding information of the SV40 early region fused to sequences upstream of the rat insulin (II) gene. The boxed regions denote insulin gene flanking DNA and the thick solid line refers to the SV40 early region. The cap site where insulin gene transcription initiates is indicated by an L-shaped arrow, and the associated transcriptional enhancer element (EH) is shown; the insulin gene promoter, which lies between them, is not indicated. The points of SV40 T-antigen translation initiation and transcription termination are indicated by boxed ATG and An, respectively.

mosaic for the acquired transgene, using the criteria of infrequent transmission to progeny and lower apparent copy number in the founder than in its progeny. The four independently derived transgenic mice are listed in Table 1, which presents their characteristic transgene copy numbers and strain designations. In each case, multiple copies of the injected DNA have been acquired. Genetic and biochemical analysis indicates that the genes are heterozygous and integrated in a single location (not shown).

Phenotype of Insulin/T-Antigen Transgenic Mice

During the initial breeding analysis of the insulin/T-antigen transgenic mice, a phenotype of premature death emerged. The two original nonmosaic mice died at 9–12 weeks of age. One of these transmitted the transgene to offspring before dying, while the other (M163) did not. The phenotype of premature death proved to be heritable, and cosegregated with the acquired hybrid gene. A partial pedigree is given in Figure 2 in order to illustrate this observation. The penetrance can be seen to be virtually complete in the early generations. The analysis which follows deals primarily with the progeny of the original nonmosaic mouse which transmitted, M164 (RIP1-Tag #2). Progeny of M164 were sacrificed and subjected to pathological and biochemical analysis. The only obvious abnormal pathology consisted of hyperplasia of the islets of Langerhans, as determined following standard histochemical staining of tissue sections. The continuing analysis of the mice in the RIP1-Tag #2 lineage indicates that all mice harboring this insertion evidence the phenotype of islet hyperplasia (if sacrificed) and/or premature death. Prior to their sudden death, the mice seem normal in appearance and activity. The pedigree, shown in Figure 2, stops at the generation during which the mice were placed on a high sugar diet to counteract indications that the mice were hypoglycemic, which is a likely consequence of islet hyperplasia, and a possible cause of sudden, premature death. Subsequent generations, maintained on this diet, now live somewhat longer, and heritably develop solid tumors of the pancreas between 10 and 20 weeks of

Table 1
Independent Mice Harboring the RIP1-Tag Oncogene

Mouse number	Genotype[a]	Heterozygous copy number	Comments
163	RIP1-Tag #1	4–5	Died without transmitting
164	RIP1-Tag #2	4–5	
165	RIP1-Tag #3	25[b]	Founder mosaic
168	RIP1-Tag #4	12[b]	Founder mosaic

[a]Each founder mouse (derived from an injected embryo) is given a genotype number to indicate its independent origin, and its presumed unique site of integration.

[b]Since the founder mouse was mosaic for the acquired transgene, the copy number is that of nonmosaic progeny.

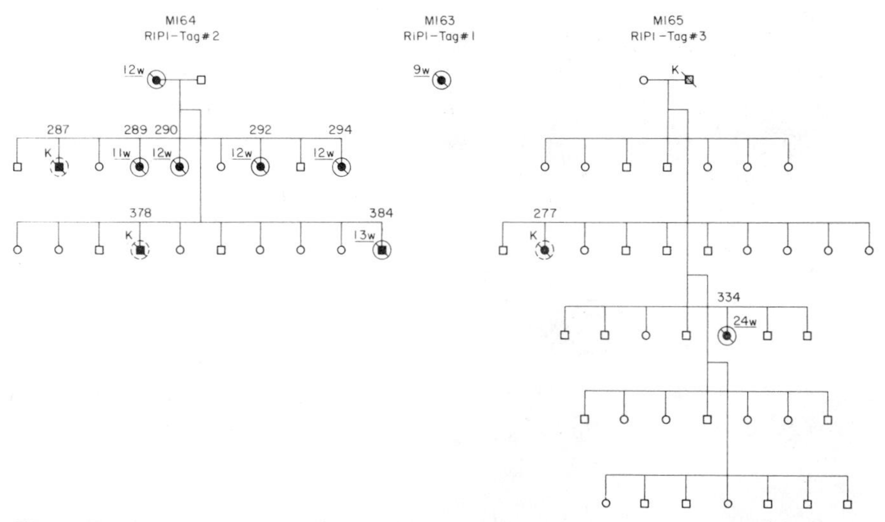

Figure 2
An early pedigree of insulin/T-antigen transgenic mice. Females are indicated by circles, males by squares, and transgenic mice by blackened circles and squares. The circled slash denotes premature death, with the age in weeks shown. A slash alone indicates the animal was sacrificed, and the dotted circle around a sacrificed mouse denotes evidence of islet hyperplasia following histopathology. Each independent insertion of a particular gene is given a number, so as to distinguish the separate lines developing from each founder mouse. Mouse 165 is mosaic both by the genetic transmission data presented here, and by lower apparent heterozygous copy number relative to progeny.

age. These tumors come to comprise a significant fraction of the mass of the pancreas, and are highly vascularized. Individual tumors can reach a size of 5 mm in diameter, with up to five or six separate tumors visible in a pancreas. There is no indication that the tumors metastacize other organs.

Cell Type Specificity of Fusion Gene Expression

To evaluate the observed hyperplasia of the islets of Langerhans, pancreas' from both normal and transgenic mice were examined for expression of SV40 T-antigen, as well as for the expression of the polypeptide hormones characteristic for the major cell types of the islets. Thin sections were prepared from fresh frozen pancreas obtained from a transgenic mouse (M384 of the RIP1-Tag #2 lineage) carrying visible pancreatic tumors, as well as from a normal mouse of similar age. The sections were analyzed by immunostaining, using antisera to glucagon, insulin, and somatostatin, each visualized with rhodamine-conjugated second antibody, in order to identify the α, β, and δ cell types of the islets. T-antigen was identified using both a polyclonal antiserum and monoclonal antibodies, each visualized by horseradish peroxidase conjugated second antibody. Figure 3, left column, shows

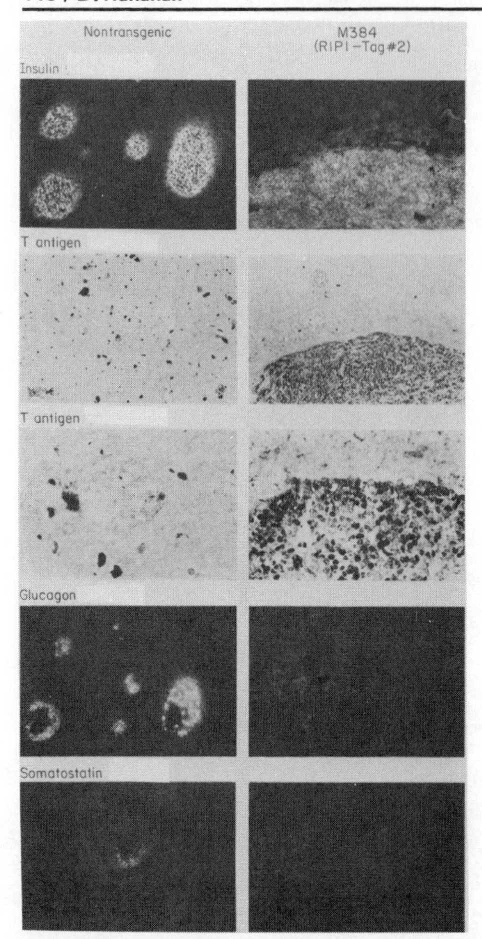

Figure 3

Analysis of islet cell hormone and T-antigen expression in a normal mouse and one tumor-bearing transgenic mouse. Each column shows an analysis of one mouse: The left column is a normal mouse, the right a mouse (384) from the RIP1-Tag #2 line. In each case nearby (but not necessarily adjacent) thin sections were analyzed for the presence of the noted antigen by using an antisera to that antigen followed by a conjugated second antibody to visualize the first. The first row examines insulin; the second and third, T-antigen; the fourth, glucagon; and the fifth, somatostatin. The magnification is 50X except for the third row, which is 200X. The use of second antibodies alone evidences no specific staining. In the normal mouse column, insulin and glucagon plates are of the same cluster of islets; but the somatostatic and T-antigen are of other islets, as that cluster was not present in those sections. The immunostains for the hormones were visualized with rhodamine-conjugated second antibody and epifluorescent illumination. The immunostain for T-antigen employed peroxidase-conjugated second antibody, and these immunostains were viewed under bright field illumination. In the middle and right columns, the orientation of all the plates is the same. The islet tumor is on the bottom, with adjacent exocrine tissue about it. The M384 column is aligned so the tumor is a centered arc. The glucagon and somatostatin plates of each are of sections near those used for insulin and T-antigen, and the outline of the exocrine/tumor boundary can be seen in each. The 200X plates of T-antigen also show an exocrine/tumor boundary, in which peroxidase staining is seen in nuclei in the tumor tissue but not in the exocrine tissue.

representative immunostains of normal (nontransgenic) islets for the presence of insulin, T-antigen, glucagon, and somatostatin. The islets can be seen to be comprised predominantly of α and β cells, with fewer δ cells, and no cells reacting with antibodies to T-antigen. Controls using second antibodies alone evidenced no specific staining (not shown).

The islet tumor (Fig. 3, right column), in contrast, appears to consist solely of insulin producing cells, which are identifiable as β cells using both immunostaining and histochemical staining. Glucagon and somatostatin producing cells are rare if not completely absent from the tumors. Characteristic localization of T-antigen to the nucleus is observed in the islet tumor cells, and it is clearly absent from adjacent exocrine cells. Both polyclonal serum (shown) and monoclonal antibodies (not shown) evidence the same pattern of T-antigen expression. By these criteria, the islet tumors are essentially pure populations of proliferating β cells. Thus the hybrid oncogene is specifically inducing proliferation of the β cells of the islets of Langerhans, eventually resulting in the appearance of pure β cell tumors. In comparison, naturally occurring insulinomas are characteristically a mixed population of both β and δ cells, which are producing insulin and somatostatin, respectively (Chick et al. 1977; Gazdar et al. 1980).

Tissue-specific Expression of Insulin-Tag Fusion Genes

The ability of the insulin control region to mediate tissue-specific expression of the SV40 T-antigen following integration of the hybrid genes into the mouse germline was assessed by protein blotting analyses (Towbin et al. 1979; Burnette 1981), using monoclonal antibodies which specifically recognize T-antigen (Harlow et al. 1981). Figure 4a presents a comparison of pancreas' from several mice, including one normal (nontransgenic) mouse, several insulin-Tag transgenic mice, and one carrying a collagen T-antigen fusion gene (M410), which therefore lacks the insulin gene sequences (D. Hanahan unpubl.). M287 and M289 are siblings (see Fig. 2). M287 had large solid tumors, whereas M289 did not, which illustrates the difference in levels of expression observed in mice of the same age and genotype, with and without solid tumors. M277 can be seen to contain immunoprecipitable T-antigen, which demonstrates that the RIP1-Tag #3 lineage derived from the founder M165 also expresses the fusion gene. T-antigen is not detectable in a normal (F1) pancreas or in a transgenic mouse (M410) which lacks the insulin gene regulatory region. M274 does not evidence expression in this analysis. Nevertheless, mice in the RIP1-Tag #4 lineage are occasionally developing β cell tumors.

The tissue specificity of fusion gene expression was further evaluated by comparing a number of different tissues from two mice (Fig. 4b,c), progeny of the two founders M164 and M165. Both show tissue-specific expression of SV40 T-antigen, as only pancreas is producing immunoprecipitable protein which comigrates with authentic T-antigen. Some degradation of T-antigen in pancreatic tissue is routinely observed; it is not clear whether this represents a normal situation in the β cells or is

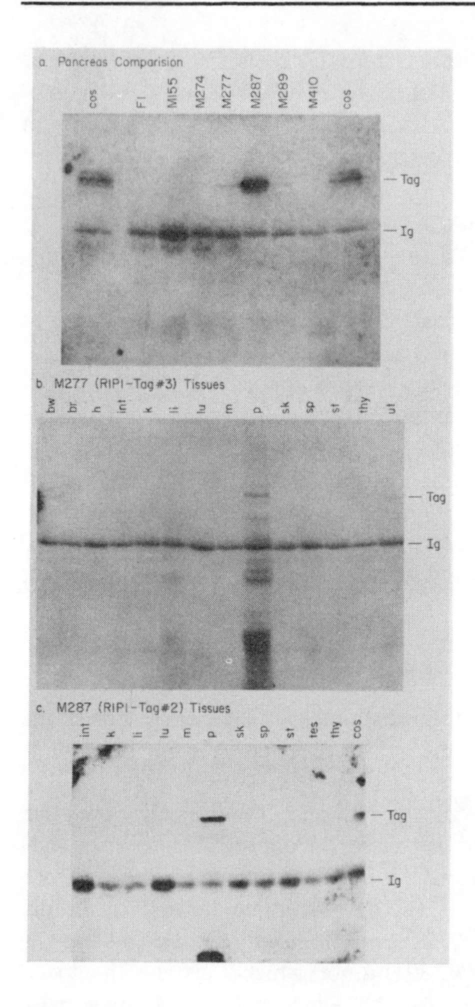

Figure 4

Tissue specificity of transgene expression. The presence of SV40 T-antigen in tissue homogenates was evaluated by immunoprecipitation with a monoclonal antibody, fractionation on an SDS polyacrylamide gel, transfer to nitrocellulose by electroblotting, and visualization with a second radiolabeled monoclonal antibody followed by autoradiography. (*a*) Pancreas from several mice are compared. F1 is a nontransgenic mouse; M274 is RIP1-Tag #4; M277 is RIP1-Tag #3. M287 and M289 are from siblings from the RIP1-Tag #2 lineage, with M287 having visible tumors and M289 only hyperplastic islets. M410 is a mouse harboring a type I collagen/SV40 T-antigen fusion gene. M155 harbors different recombinant insulin-T-antigen gene. The T-antigen producing cell line cos A2 (Gluzman 1981) was used as a control, at a level of 5% total cos cell protein to total tissue protein. (*b,c*) Tissue specificity of Tag expression in the RIP1-Tag #2 and #3 lineages. Equal amounts of each tissue were examined. The tissue codes are as follows: (bw) body wall; (br) brain; (h) heart; (int) intestine; (k) kidney; (li) liver; (lu) lung; (m) muscle; (p) pancreas; (sk) skin; (st) stomach; (tes) testes; (thy) thymus; (ut) uterus. A separate analysis indicates that no T-antigen can be detected in body wall, brain, or heart of M287 (not shown).

simply a consequence of the protein isolation procedure. Analysis of the tissues of the founder mouse 163 also shows expression of T-antigen exclusively in the pancreas (not shown). Taken together with the immunohistochemical analysis, the transferred insulin sequences are specifying correct expression in the β cells of the endocrine pancreas.

Penetrance and the Onset of Oncogenesis

No more than four or five of the approximately 100 islets in a pancreas develop into solid β cell tumors, which are highly vascularized and readily distinguishable from normal pancreatic tissue. Yet most islets evidence hyperplasia; all islets examined have increased β cell density. Furthermore, the β cells appear to be expressing the hybrid genes well before the development of hyperplasia and tumors. This is demonstrated in an examination of the pancreas from an 8-week-old mouse (M633) in the RIP1-Tag #2 lineage. This mouse exhibited normal pathology and histology, in particular showing no hyperplasia or visible tumors in the pancreas. Thin sections were prepared from the pancreas, and stained with antibodies for insulin and T-antigen. Figure 5 shows a representative analysis of a pair of islets. T-antigen is expressing in both islets shown, as well as in all other islets examined in this pancreas. Thus T-antigen is expressing in the β cells of a young mouse, prior to emergence of obvious hyperplasia and solid tumors.

DISCUSSION

The results presented here demonstrate that DNA within 660 bp upstream of the rat insulin (II) gene is sufficient to mediate qualitatively correct tissue and cell type specific expression of hybrid insulin/SV40 T-antigen genes. Protein blotting analysis of mice harboring independent insertions of such genes evidences expression only

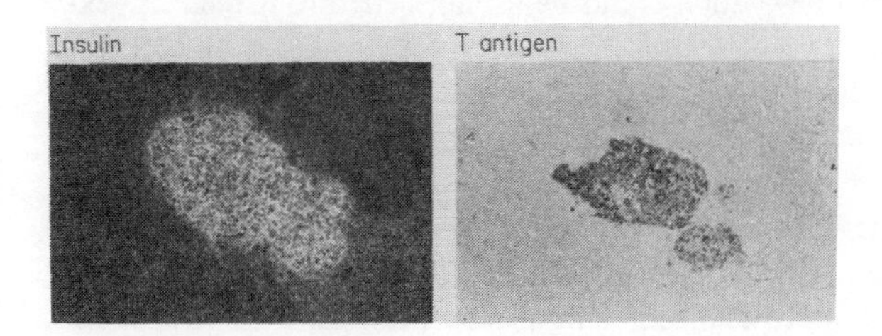

Figure 5
T-antigen expression in the islets of a young transgenic mouse. Mouse 633, from the RIP1-Tag #2 lineage, was sacrificed at 8 weeks of age, and its pancreas was flash frozen in liquid nitrogen, sectioned in a cryostat, and analyzed for the expression of islet cell hormones and T-antigen as described in Figure 3. The two islets shown are representative of all islets in these and other thin sections from this pancreas. The sections used for insulin on T-antigen are nearby but not adjacent, and the islet sizes have changed somewhat across that distance.

in the pancreas among the major tissues examined. The in situ analyses of thin sections from the pancreas demonstrates that this expression is cell type specific, as T-antigen is only detected in the insulin producing β cells of the islets of Langerhans.

These observations represent a step toward identifying those *cis*-acting elements necessary to elicit tissue-specific expression of the insulin genes. The localization of these elements to a 660 bp region that encompasses the promoter and enhancer for transcriptional initiation suggest their involvement in tissue specificity. Dissection of this region so as to attempt separation of the tissue specificity from promoter and/or enhancer should provide more information on the location, and perhaps functional manifestation, of tissue specificity. Work on immunoglobulin gene regulation suggests that the promoter element and perhaps the gene body may be important for tissue specificity, in addition to the observed specificity of the transcriptional enhancer element (R. Grosschedl and D. Baltimore, in prep.).

Two other qualities of insulin gene regulation remain to be examined for the hybrid insulin-SV40 gene. The endogeneous insulin genes are activated at the fourteenth to fifteenth day of development in mouse and rat (Pictet and Rutter 1972; McEvoy and Madson 1980; Cooperstein and Watkins 1981; Kakita et al. 1983). The time that the transferred gene is switched on will indicate whether the 5′ flanking region of the insulin gene carries this temporal information as well. The insulin genes are transcriptionally induced by glucose, both in vivo and in primary cultures of β cells (Brunstedt and Chan 1982; Kakita et al. 1982, 1983). The culture of the transgenic β cells in medium containing inducing and noninducing amounts of glucose should allow an assessment of the inducibility of the hybrid insulin gene, using the induction of the endogenous genes as an internal standard. The evaluation of these two characteristics will also motivate dissection of the 5′ flanking region if both are correctly mediated by this region, or addition of other portions of the insulin gene if they are not. It remains formally possible that temporal regulation and/or glucose inducibility will prove species-specific—although tissue specificity seems not to be—which could be established through similar experiments utilizing the mouse flanking region.

An additional goal of utilizing hybrid oncogenes in transgenic mice is to use their cell-specific expression to expand rare cell types into established lines of cultured cells that are representative of those cells. Isolation of β cell lines from these transgenic mice should be facilitated by the ability of T-antigen to establish (or immortalize) primary cells to growth in culture, with T-antigen synthesis here maintained through the coexpression of the insulin and insulin/T-antigen genes in β cells. The availability of β cell lines will allow more detailed examination of the characteristics of the transgenic β cells. An important criteria for an authentic β cell is the continued ability of its insulin genes to be induced by glucose. This quality is lost in cultured insulinoma cells, although it is usually present in the tumors from which they are derived (Gazdar et al. 1980). The use of an appropriate insulin-oncogene in transgenic mice may allow immortalization without loss of inducibility.

The hybrid insulin/SV40 T-antigen gene is apparently expressed in all β cells, thereby inducing a general proliferative state which is manifested as a hyperplasia of the islets of Langerhans. Solid β cell tumors eventually develop from a few of the hyperplastic islets. Tumor development occurs between 10 to 20 weeks of age, is heritable, and has occurred in all progeny of the RIP1-Tag #2 line that have lived to this age, and is now occurring in the RIP1-Tag #3 line. The solid tumors are frequent in that they arise in every mouse harboring these genes, although only in a fraction of the islets. The characteristics of tumor formation suggest that synthesis of T-antigen is necessary but insufficient to produce oncogenesis in every islet. T-antigen is of a class of oncogenes that can both immortalize primary cells and release (or transform) cultured cells from their normal growth controls, such as contact inhibition and dependence on serum growth factors (Tooze 1981). Yet, in vivo, SV40 virus will produce tumors in mice only after a long latent period, and oncogenesis appears to be related to the efficacy of the immune response and the presence of T-antigen (TSTA) on the cell surface as well as in the nucleus (Hargis and Malkiel 1979; Abramczuk et al. 1984). There is no evidence of an immune response to hyperplastic islets, nor of T-antigen on the surface of β cells, and it remains to be established whether the frequency of solid tumor formation is associated with an escape from immune surveillance. There are other potential alterations that might be necessary to elicit the development of solid β cell tumors. These include activation or inactivation of growth factors or receptors, or the cooperation of another oncogene (Land et al. 1983; Ruley 1983). It is well established that nascent tumors cannot exceed certain size constraints without becoming vascularized, and thus the induction of angiogenesis is likely to be one secondary event necessary for solid tumor development (Folkman et al. 1971).

The oncogenic conversion event which leads to solid tumor formation can be loosely classified as either a cell heritable mutation, which converts a single β cell into a tumor cell, or a general islet alteration, that results in most of the β cells in an islet participating in tumor formation. Distinguishing between these possibilities is important for designing experiments to elucidate the mechanisms of this event, which could well bear on the multistep theories of cancer. A cell heritable mutation would motivate attempts to isolate and study that mutation. These possibilities can in part be distinguished through determination of the clonality of individual islet tumors. One possible way to do so involves the assessment of X inactivation using distinguishable markers on the two X chromosomes of female mice. If the tumor is clonal, it should contain only one active X, whereas a nonclonal tumor will contain both forms of a polymorphic marker. This approach, however, assumes that individual islets are not already clonal.

Another approach to the mechanism of oncogenesis involves the introduction of a series of recombinant insulin-oncogenes, using structural information for such oncogenic proteins as *myc, ras, fos, E1A,* and *p53*. The phenotype they elicit may provide more information on their common and distinguishable properties. Cooperation experiments can then be performed by mating transgenic mice harboring different recombinant oncogenes, so as to examine the development of oncogenesis and the possibly separable roles these various proteins might play in the process.

REFERENCES

Abramczuk, J., S. Pan, G. Maul, and B.B. Knowles. 1984. Tumor induction by Simian Virus 40 in mice is controlled by long term persistance of the viral genome and the immune response of the host. *J. Virol.* **49**: 540.

Brinster, R., R.A. Kindred, R.E. Hammer, R.L. O'Brien, B. Arp, and U. Storb. 1983. Expression of a microinjected immunoglobulin gene in the spleen of transgenic mice. *Nature* **306**: 332.

Brinster, R. L., H.Y. Chen, A. Messing, T. van Dyke, A.J. Levine, and R.D. Palmiter. 1984. Transgenic mice harboring SV40 T-antigen genes develop characteristic brain tumors. *Cell* **37**: 367.

Brunstedt, J. and S.J. Chan. 1982. Direct effect of glucose on the preproinsulin mRNA level in isolated pancreatic islets. *Biochem. Biophys. Res. Commun.* **106**: 1383.

Burnette, W.N. 1981. Western blotting: Electrophoretic transfer of proteins from SDS-polyacrylamide gels to unmodified nitrocellulose and radiographic detection with antibody and radioiodinated protein A. *Anal. Biochem.* **112**: 195.

Chada, K., J. Magram, K. Raphael, G. Radice, E. Lacy, and F. Costantini. 1985. A foreign β-globin gene expressed specifically in erythroid cells of transgenic mice. *Nature* (in press).

Charnay, P., R. Treisman, P. Mellon, M. Chao, R. Axel, and T. Maniatis. 1984. Differences in human α- and β-globin gene expression in mouse erythroleukemia cells: The role of intragenic sequences. *Cell* **38**: 251.

Chick, W.L., S. Warren, R.N. Chute, A.A. Like, V. Laurig, and K.C. Kitchen. 1977. A transplantable insulinoma in the rat. *Proc. Natl. Acad. Sci. U.S.A.* **74**: 628.

Cooperstein, S.J. and D. Watkins, eds. 1981. *The Islets of Langerhans. Biochemistry, physiology, and pathology.* Academic Press, New York.

Cordell, B., G. Bell, E. Tischer, F. DeNoto, A. Ullrich, R. Pictet, W. Rotter, and H. Goodman. 1979. Isolation and characterization of a cloned rat insulin gene. *Cell* **18**: 533.

Folkman, J., E. Merler, C. Abernathy, and G. Williams. 1971. Isolation of a tumor factor responsible for angiogenesis. *J. Exp. Med.* **133**: 275.

Gazdar, A., W.L. Chick, H.K. Oie, H.L. Sims, D.L. King, G.C. Wein, and V. Lauris. 1980. Continuous, clonal, insulin- and somatostatin-secreting cell lines established from a transplantable rat islet cell tumor. *Proc. Natl. Acad. Sci. U.S.A.* **77**: 3519.

Gluzman, Y. 1981. SV40-transformed simian cells support the replication of early SV40 mutants. *Cell* **23**: 175.

Gordon, J.W., G.A. Scangos, D.J. Plotkin, J.A. Barbosa, and F.H. Ruddle. 1980. Genetic transformation of mouse embryos by microinjection of purified DNA. *Proc. Natl. Acad. Sci. U.S.A.* **77**: 7380.

Grosschedl, R., D. Weaver, D. Baltimore, and F. Costantini. 1984. Introduction of a μ immunoglobulin gene into the mouse germ line: Specific expression in lymphoid cells and synthesis of functional antibody. *Cell* **38**: 647.

Hargis, B.J. and S. Malkiel. 1979. Sarcomas induced by injection of Simian Virus 40 into neonatal CFW mice. *J. Natl. Cancer Inst.* **63**: 965.

Harlow, E., L.V. Crawford, D.C. Pim, and N.M. Williamson. 1981. Monoclonal antibodies specific for Simian Virus 40 tumor antigens. *J. Virol.* **39**: 861.

Kakita, K., S. Goddings, and M. Permutt. 1982. Biosynthesis of rat insulins I and II: Evidence for differential expression of the two genes. *Proc. Natl. Acad. Sci. U.S.A.* **79**: 2803.

Kakita, K., S.J. Giddings, P.S. Rotwein, and M.A. Permutt. 1983. Insulin gene expression in the developing rat pancreas. *Diabetes* **32**: 691.

Lacy, E., S. Roberts, E.P. Evans, M.D. Burtenshaw, and F.D. Costantini. 1983. A foreign β-globin gene in transgenic mice: Integration at abnormal chromosomal positions and expression in inappropriate tissues. *Cell* **34**: 343.

Land, H., L.F. Parada, and R.A. Weinberg. 1983. Tumorigenic conversion of primary embryo fibroblasts requires at least two cooperating oncogenes. *Nature* **304**: 596.

Lomedico, P., M. Rosenthal, A. Efstratiadis, W. Gilbert, R. Kolodner, and R. Tizard. 1979. The structure and evolution of the two nonallelic rat preproinsulin genes. *Cell* **18**: 545.

McEvoy, R.C. and K.L. Madson. 1980. Pancreatic insulin-, glucagon-, and somatostatin-positive islet cell populations during the perinatal development of the rat. *Biol. Neonate* **38**: 248.

McKnight, G.S., R.E. Hammer, E.A. Kuenzel, and R.L. Brinster. 1983. Expression of the chicken transferrin gene in transgenic mice. *Cell* **34**: 335.

Palmiter, R., H. Chen, and R. Brinster. 1982a. Differential regulation of metallothionein-thymidine kinase fusion genes in transgenic mice and their offspring. *Cell* **29**: 701.

Palmiter, R.D., R.L. Brinster, R.E. Hammer, M.E. Trumbauer, M.G. Rosenfeld, N.C. Birnberg, and R.M. Evans. 1982b. Dramatic growth of mice that develop from eggs microinjected with metallothionein-growth hormone fusion genes. *Nature* **300**: 611.

Pictet, R. and W.J. Rutter. 1972. Endocrine pancreas. In *Handbook of physiology* (ed. D.F. Steiner and N. Frienkel), vol. I, sec. 7, p. 25. Williams and Wilkins, Baltimore, Maryland.

Ruley, H.E. 1983. Adenovirus early region 1A enables viral and cellular transforming genes to transform primary cells in culture. *Nature* **304**: 602.

Steiner, D.F. and N. Freinkel, eds. 1972. Endocrine pancreas. In *Handbook of physiology*, vol. 1, sec. 7. Williams and Wilkins, Baltimore, Maryland.

Stewart, T.A., P.K. Pattengale, and P. Leder. 1984. Spontaneous mammary adenocarcinomas in transgenic mice that carry and express MTV/myc fusion genes. *Cell* **38**: 627.

Swift, G.H., R.E. Hammer, R.J. MacDonald, and R.L. Brinster. 1984. Tissue-specific expression of the rat pancreatic elastase I gene in transgenic mice. *Cell* **38**: 639.

Tooze, J., ed. 1981. *Molecular biology of tumor viruses Pt. 2 (revised)*. Cold Spring Harbor Laboratory, Cold Spring Harbor, New York.

Towbin, H., T. Staehelin, and J. Gordon. 1979. Electrophoretic transfer of proteins from polyacrylamide gels to nitrocellulose sheets: Procedure and some applications. *Proc. Natl. Acad. Sci. U.S.A.* **76**: 4350.

Ullrich, A., J. Shine, J. Chirgwin, R. Pictet, E. Tischer, W. Rutter, and H. Goodman. 1977. Rat insulin genes: Construction of plasmids containing the coding sequences. *Science* **196**: 1313.

Villa-Komaroff, L., A. Efstratiadis, S. Broome, P. Lomedico, R. Tizard, S.P. Naber, W. L. Chick, and W. Gilbert. 1978. A bacterial clone synthesizing proinsulin. *Proc. Natl. Acad. Sci. U.S.A.* **75**: 3727.

Walker, M.D., T. Edlund, A.M. Boulet, and W.J. Rutter. 1983. Cell-specific expression controlled by the 5′-flanking region of insulin and chymotrypsin genes. *Nature* **306**: 557.

Wright, S., A. Rosenthal, R. Flavell, and F. Grosveld. 1984. DNA sequences required for regulated expression of β-globin genes in murine erythroleukemia cells. *Cell* **38**: 265.

Transgenic Mice Carrying MMTV myc Fusion Genes Develop Mammary Adenocarcinomas

TIMOTHY STEWART,*[1] PAUL PATTENGALE,† AND PHILIP LEDER*
*Department of Genetics
Harvard Medical School
Boston, Massachusetts 02115
†Department of Pathology
University of Southern California School of Medicine
Los Angeles, California 90033

OVERVIEW

Alterations in the *myc* gene have been considered to be one of the steps in the process of cellular transformation. To provide a means of analyzing the in vivo consequences of aberrant *myc* expression, 13 lines of mice have been produced by pronuclear injection of artificially constructed c-*myc* genes. Each myc mouse carries a section of the *myc* gene that has been fused to the hormone-inducible mouse mammary tumor virus long terminal repeat (MMTV LTR). In at least 11 of the 13 transgenic mice, transcripts of the newly introduced *myc* gene are present in one or another of the mouse organs. Although development is apparently normal in the heterozygous offspring of these animals, mammary adenocarcinomas developed in the founder animals of two of these lines. The tumors appeared during pregnancy in multipared females at about the age of 7 months; in each case the tumor contained multiple copies of the newly introduced *myc* gene and its transcript. Similarly, all available transgenic F1 daughters of one of these founders developed a breast tumor during their second or third pregnancy. This result is consistent with the participation of the newly introduced germline MMTV *myc* gene in development of these malignancies in transgenic animals. These mouse lines provide model systems for assessing factors that influence tumor formation in animals carrying an activated oncogene as a part of their constitutional genome.

INTRODUCTION

The alterations in the growth characteristics that accompany the cellular transition from a normal pattern to a neoplastic state seem to be a result of changes in a relatively small number of proto-oncogenes (Cairns 1981; Klein 1981; Mintz and Fleischman 1981; Bishop 1983; Land et al. 1983a; Leder et al. 1983). The genetic alterations in these genes include promoter insertions, amplifications, somatic mu-

[1]*Present address*: Department of Molecular Biology, Genentech, Inc., 460 Pt. San Bruno Boulevard, South San Francisco, California 94080

tations, and chromosomal translocations. In the case of the *myc* gene, all of these have been observed in one or more tumor types (Leder et al. 1983). The possibility that the relationship between aberrant *myc* expression and tumorigenesis may be more than casual has been recently supported by the observation that the transfection of a combination of tumor-derived *ras* and *myc* genes into primary rat fibroblasts produces cellular transformation, whereas transfection of either alone does not (Land et al. 1983b).

To begin an examination of the relationship between aberrant *myc* expression and in vivo tumorigenesis, we produced a series of transgenic mice carrying in their germline a *myc* gene that was fused to the mouse mammary tumor virus long terminal repeat (MMTV LTR). Transgenic mice have been used to examine a variety of genes and their regulation (Brinster et al. 1981; Costantini and Lacy 1981; Wagner et al. 1981; Palmiter et al. 1982; Stewart et al. 1982; Brinster et al. 1983; Lacy et al. 1983; McKnight et al. 1983; Palmiter et al. 1983). The transgenic animals described here should allow us to assess the effect of a hormonally-inducible oncogene, present in each cell of the animal, on the normal processes of development and differentiation. Furthermore, they should allow us to determine whether this activated oncogene can collaborate with a normally expressed gene to produce malignant transformation or whether the somatic mutation of an additional oncogene (or oncogenes) is required for tumorigenesis. In the results we describe here, we show that the presence and transcription of the MMTV *myc* gene does not grossly interfere with normal development in heterozygous animals, but – in at least two lines – does appear to predispose to malignancies, particularly those of the breast.

RESULTS

Production of 13 Lines of Transgenic Mice

To examine the consequences of abnormal *myc* activity in an in vivo context, 13 lines of transgenic mice that carry, integrated into the genome, one of four different MMTV *myc* constructions have been generated. The MMTV *myc* genes were produced by a series of fusions between the MMTV LTR and the *myc* gene (Figure 1). The MMTV LTR (including the region required for glucocorticoid control [Ringold 1983], the MMTV promoter and cap site) was taken from the PA9 plasmid described by Huang et al. (1981) and a subclone of the mouse *myc* gene (Battey et al. 1983) provided the *myc* regions. The details of the fusions have been published elsewhere (Stewart et al. 1984).

The four MMTV *myc* gene plasmids were digested with *Sal*I and *Eco*RI (each of which cleaves once within the pBR322 sequence) and separately injected into the male pronuclei of fertilized one cell mouse eggs. The transgenic founder animals that resulted from the injected eggs were used to establish the 13 lines of mice shown in Table 1. Each line carries the indicated MMTV *myc* construction at apparently a single chromosomal locus.

A. STARTING PLASMIDS

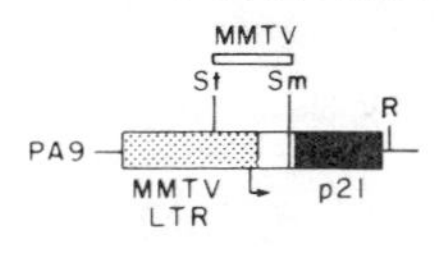

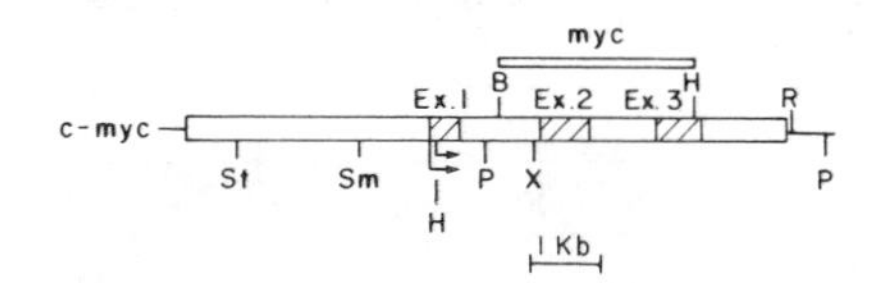

B. FUSION GENES

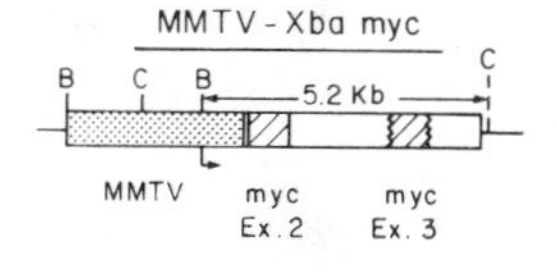

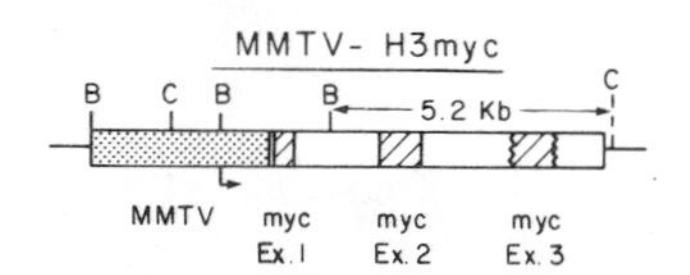

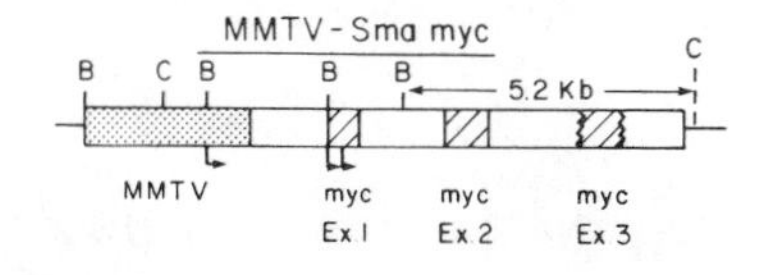

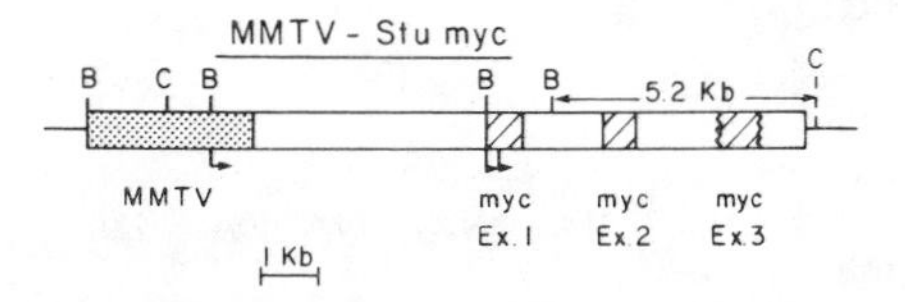

Figure 1
(*A*) Structure of the MMTV and *myc* containing plasmids used to construct the MMTV *myc* fusion genes. Plasmid pA9 is the MMTV LTR containing vector (Huang et al. 1981) used as the source of the MMTV LTR sequences in the MMTV *myc* constructions described here and the *myc* sequences were derived from a subclone of the mouse *myc* gene cloned from an embryonic mouse library (Battey et al. 1983). (*B*) The MMTV myc fusion genes. The restriction sites are *Stu*I (St); *Sma*I (Sm); *Eco*RI (R); *Hin*dIII (H); *Pvu*I (P); *Bam*HI (B); *Xba*I (X); *Cla*I (C) (not all sites are marked). The solid arrows below the constructions represent the promoter in the MMTV LTR and in the *myc* gene. The size (in kb) of the major fragment produced by digestion with *Bam*HI and *Cla*I is shown for each construction. The two probes (*myc* and MMTV) used for the majority of the filter hybridization analyses are shown. Details of the construction of these MMTV *myc* genes are given in Stewart et al. (1984). Reprinted, with permission, from Stewart et al. (1984).

So as to be able to correlate any phenotype seen in these MMTV myc mice with the activity of the MMTV *myc* genes, we have analyzed the level and extent of expression of these newly introduced hormone responsive genes. Because transcription initiated at the MMTV promoter will transcribe across the fusion point between the *myc* and the MMTV sequences, we have used DNA fragments (cloned into M13 vectors) that span this fusion point in S1 nuclease protection assays. As the probes used include *myc* sequences, they will detect expression of the endogenous mouse *myc* genes, but will also protect unique fragments that are diagnostic for transcription of each of the MMTV *myc* genes. Details of the probes and procedures have been published elsewhere (Stewart et al. 1984).

A summary of the data is shown in Table 1. Of the 12 lines in which organs were examined for expression, novel transcripts were detected in the salivary glands of 11, and in three of these lines the salivary glands were the only organ in which these transcripts were seen. In one case, 147-9a, the MMTV *Sma myc* gene was expressed

Table 1
Novel Transcripts of MMTV *myc* Fusion Genes in Various Organs of Transgenic myc Mice

	Source of RNA													
myc Mice	Th	Pa	Sp	Ki	Te	Li	He	Lu	Mu	Br	Sa	Pr	In	Ma
MTV-Stu														
155-2	–	–	–	–	–	–	–	–	–	–	+	–	NA	NA
155-7	–	–	–	–	–	–	–	–	–	–	+	–	NA	NA
MTV-Sma														
147-6	–	–	–	–	–	–	–	–	–	–	+	–	NA	NA
147-9a	+	+	+	+	+	+	ND	+	+	+	+	+	NA	NA
147-9b	–	–	+	–	+	–	–	+	–	+	+	+	NA	NA
MTV-H3														
139-1	–	–	–	–	+	–	–	+	–	+	+	–	NA	NA
141-3	–	–	–	–	–	–	–	–	–	–	+	–	+	+
143--3	–	–	–	–	–	–	–	–	–	–	+	–	NA	NA
143-5	–	–	–	–	–	–	–	–	–	–	–	–	–	–
MTV-Xba														
164-4a	NA	NA	NA	NA	NA	NA	NA	NA	NA	NA	NA	NA	NA	NA
164-4b	+	–	–	–	–	–	–	+	–	–	+	–	+	+
165-1a	–	–	–	–	+	–	–	–	–	–	+	–	NA	NA
165-1b	–	–	–	–	+	–	–	–	–	–	+	–	NA	NA

Abbreviations and Symbols: (Th) Thymus; (Pa) pancreas; (Sp) spleen; (Ki) kidney; (Te) testis; (Li) liver; (He) heart; (Lu) lung; (Mu) muscle (striated); (Br) brain; (Sa) salivary gland; (Pr) preputial gland; (In) intestine (small); (Ma) mammary gland; (NA) not assayed; (+) MMTV *myc* fusion gene transcript present; (–) MMTV *myc* fusion gene transcript undetectable.

All assays were carried out according to the S1 nuclease protocol described under Methods and in the legend to Figure 3. Reprinted, with permission, from Stewart et al. (1984).

in each organ examined. In another line, 143-5, no novel transcripts were seen. As females are being used to generate animals homozygous for these genes, we have not yet been able to examine mammary tissue in all strains. However, in two of the three lines examined, novel transcripts were observed (lines 141-3, 164-4b; Table 1).

Appearance of Spontaneous Mammary Adenocarcinomas in Two MMTV myc Transgenic Mice

Two of the founder transgenic mice that initiated the lines of MMTV myc mice were noted to have grossly obvious tumor-like masses: One (141-3; MMTV H3*myc*) developed a subcutaneous mass in the right neck region at approximately 206 days of age and the other (164-4; MTV *Xba myc*) developed a subcutaneous mass in the left groin area at 228 days of age. Significantly, both of these mice were female and both were pregnant at the time of diagnosis. Biopsies of the original growths were obtained and the tissues were fixed, stained, and examined histologically. Both primary tumors were diagnosed as moderately well-differentiated mammary adenocarcinomas arising in and involving breast tissue (Fig. 2A). The tumor in mouse 141-3 recurred approximately 3 weeks later (again in the latter stages of pregnancy) with bilateral masses in the neck and groin. At autopsy, metastatic tumor was found in the lungs (Fig. 2B). Although moderately well-differentiated in terms of gland formation, the neoplastic epithelial cells demonstrated a considerable degree of nuclear atypia (Fig. 2C) accompanied by a high mitotic rate. The tumor in mouse 164-4 recurred approximately 6 weeks later and, at autopsy, was found to be locally invasive in both subcutaneous fat and skeletal muscle. No distant metastases were observed. Salivary glands from both mice were grossly and histologically normal.

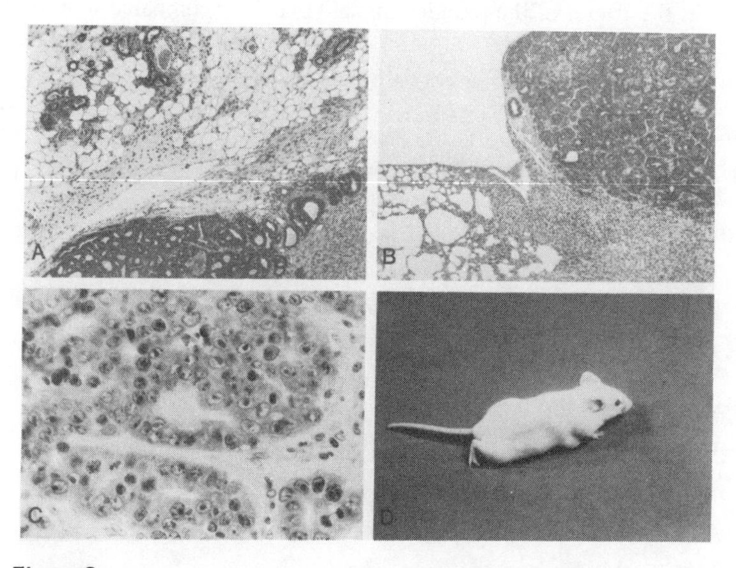

Figure 2
Histologic appearance of spontaneous mammary tumor occurring in founder mouse 141-3 and gross morphologic appearance of tumor in female progeny, 141-3-38. Virtually identical microscopic histologies were observed in founder mouse 164-4 and progeny 141-3-38. (*A*) Low power view of primary tumor demonstrating a moderately well-differentiated, locally invasive adenocarcinoma arising and involving subcutaneous mammary tissue. Note the residual, uninvolved, non-neoplastic breast tissue. (*B*) Lower power view of lung demonstrating a large subpleural nodule of metastatic adenocarcinoma. (*C*) High power view of primary mammary tumor demonstrating malignant epithelial glands. Note the piled and heaped-up appearance of the dysplastic epithelial cells which further exhibit nuclear atypia, prominent large nucleoli, and a high mitotic rate. (Mouse 141-3; H and E; [*A*], subcutaneous mammary tissue, ×70; [*B*], lung, ×70; [*C*], subcutaneous mammary tissue, ×640.) (*D*) Female offspring of 141-3 founder animal (141-3-38) at 131 days. Note tumor mass in right axillary nipple line. Animal had delivered one litter and was pregnant for the second time. Reprinted, with permission, from Stewart et al. (1984).

Subsequently, three F1 transgenic offspring of mouse 141-3 were bred to control males and during the later stages of either their second or third litters (Fig. 3), all three developed tumor masses in the nipple line suggestive of mammary adenocarcinomas. The tumors from all of these animals have been biopsied and histological examination confirms that they are also mammary adenocarcinoma (Fig. 2).

The possibility that an adventitious infection of these animals by MMTV has led to the development of these mammary adenocarcinomas is considered unlikely as the mouse strains used in these experiments (C57BL/6J and CD-1) both have a very low incidence of spontaneous mammary adenocarcinomas (Percy and Jonas 1971), and as we have not seen any mammary tumors arising in any other transgenic animals (now numbering in the hundreds) that carry unrelated newly introduced genes.

As a further test to rule out the involvement of MTV in the etiology of these adenocarcinomas, we have examined the Southern hybridization pattern of the MMTV proviruses in the tumor of the founder animal (141-3) as compared to non-

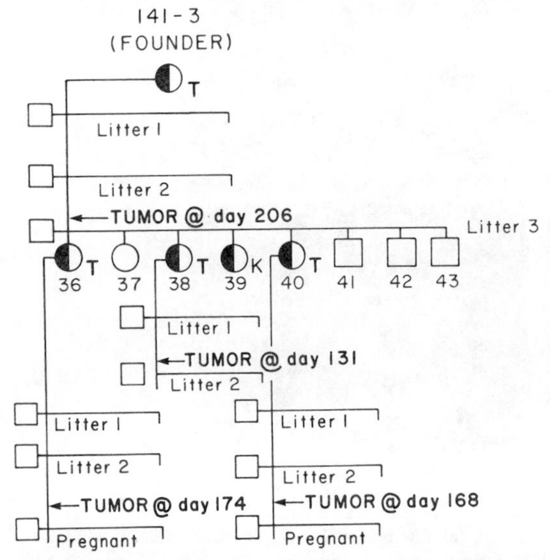

Figure 3
The occurrence of mammary adenocarcinoma among the progeny of transgenic founder 141-3. Squares and circles refer to males and females, respectively. Half-filled symbols refer to animals carrying the transgenic construction MTV-H3 *myc* on a single chromosome. The letter *T* beside a symbol indicates that the animal (founder, 36, 38, 40) developed a breast tumor; *K* indicates that the animal (39) was killed for analyses prior to any pregnancy. Litters 1 and 2 of the founder animal did not survive the initial postpartum period due to maternal neglect. The surviving litter (3) was raised by fostering to a CD1, outbred female. In most cases, including the analyzed litter 3, the male parent was an outbred CD1 mouse. The wild type progeny of this litter (37, 41, 42, 43) were not kept for analysis. Reprinted, with permission, from Stewart et. al. (1984).

tumorous tail tissue. The results indicate that neither of the MMTV sequences present endogenously or added as part of the MTV *myc* constructions have undergone amplification or gross rearrangement. We have also shown that the *myc* sequences themselves have not undergone any rearrangement (data not shown).

We have analyzed the tumors taken from both founder animals (141-3 and 164-4) for expression of the MMTV *myc* genes using S1 nuclease protection assays. Utilizing the same probes used to obtain the results shown in Table 1, we have demonstrated that in both cases the respective MMTV *myc* genes are active with transcripts approximately 20-30 times more abundant than the transcripts arising from the endogenous *myc* genes (data not shown; see Stewart et al. 1984). This result has been corroborated using a probe that spans the transcription start site in the MTV LTR.

DISCUSSION

These experiments were designed with the following points in mind: First, the effect of a constitutively active oncogene on the normal development of an adult animal could be examined; second, the pattern of expression of the MMTV *myc* genes with respect to sex, age, tissue, and chromosomal location would allow us to investigate the control of MMTV transcription in the absence of viral influence; and third, the animals carrying the active oncogene would allow an assessment of the degree to which this inherited genotype will predispose the animal to tumorigenesis and, conversely, those factors that may reverse or slow this process.

That we were able to develop 13 lines of transgenic MMTV myc mice, at least 11 of which express the newly introduced genes in one or more tissues, answers the first point. It must be noted, however, that we have not examined the activity of these genes during prenatal development when it may be that aberrant oncogene expression could be fatal.

Studies relating to the control of the MMTV LTR in these animals is at a preliminary stage; however, it is intriguing that all but one line examined express these genes in the salivary glands. Experiments are in progress to determine whether the MMTV LTR is responding to one of the hormones known to be present in high levels in salivary glands.

The occurrence of malignant mammary adenocarcinomas in two of the founder transgenic animals and in all of the F1 transgenic females of one of these founders suggests that indeed the presence and activity of the MMTV *myc* gene can lead to an increased susceptibility to tumorigenesis. Further extensive pedigree analysis will be required to demonstrate conclusively that the phenotype of susceptibility to mammary adenocarcinomas segregates with the MMTV *myc* genes. These initial results, suggesting the active MMTV *myc* genes lead to malignant conversion, also demonstrate that this one event is not sufficient. This fact is demonstrated by the observation that tumorigenesis does not involve the entire mammary epithelium.

Presumably a subsequent, somatic event (possibly a mutation) occurs in the antecedents to the tumors. Significantly it also appears that at least a third criterion, probably hormonal, must be satisfied for overt tumor growth. This conclusion is suggested by the temporal relationship between rapid tumor growth and late pregnancy.

ACKNOWLEDGMENTS

We would like to acknowledge the excellent technical assistance of Ann Kuo. We are also grateful to Dr. Gordon Hager who provided us with the MMTV hybrid plasmid pA9 and to Dr. Charles Frith who independently reviewed the histopathology. This work was supported in part by grants from the American Business for Cancer Research Foundation and E.I. DuPont de Nemours and Co., Inc. P.K.P. is a Scholar of The Leukemia Society of America.

REFERENCES

Battey, J., C. Moulding, R. Taub, W. Murphy, T. Stewart, H. Potter, G. Lenoir, and P. Leder. 1983. The human c-myc oncogene: Structural consequences of translocation into the IgH locus in Burkitt lymphoma. *Cell* **34**: 779.

Bishop, J.M. 1983. Cancer genes come of age. *Cell* **32**: 1018.

Brinster, R.L., K.A. Ritchie, R.E. Hammer, R.L. O'Brien, B. Arp, and U. Storb. 1983. Expression of a microinjected immunoglobulin gene in the spleen of transgenic mice. *Nature* **306**: 332.

Brinster, R.L., H.Y. Chen, A. Messing, T. van Dyke, A.J. Levine, and R.D. Palmiter. 1984. Transgenic mice harboring SV40 T-antigen genes develop characteristic tumors. *Cell* **37**: 367.

Cairns, J. 1981. The origin of human cancers. *Nature* **289**: 353.

Costantini, F. and E. Lacy. 1981. Introduction of a rabbit beta globin gene into the mouse germ line. *Nature* **294**: 92.

Huang, A.L., M.C. Ostrowski, O. Bernard, and G.L. Hager. 1981. Glucocorticoid regulation of the HaMuSv p21 gene conferred by sequences from mouse mammary tumor virus. *Cell* **27**: 245.

Klein, G. 1981. The role of gene dosage and genetic transposition in carcinogenesis. *Nature* **294**: 313.

Lacy, E., S. Roberts, E.P. Evans, M.D. Burtenshaw, and F.D. Costantini. 1983. A foreign beta globin gene in transgenic mice: Integration at abnormal chromosome positions and expression in inappropriate tissues. *Cell* **34**: 343.

Land, H., L.F. Parada, and R.A. Weinberg. 1983a. Cellular oncogenes and multistep carcinogenesis. *Science* **222**: 771.

———. 1983b. Tumorigenic conversion of primary embryo fibroblasts requires at least two cooperating oncogenes. *Nature* **304**: 596.

Leder, P., J. Battey, G. Lenoir, C. Moulding, W. Murphy, H. Potter, T. Stewart, and R. Taub. 1983. Translocations among antibody genes in human cancer. *Science* **222**: 765.

McKnight, G.S., R.E. Hammer, E.A. Kuenzel, and R.L. Brinster. 1983. Expression of the chicken transferrin genin transgenic mice. *Cell* **34**: 335.

Mintz, B. and R.A. Fleischman. 1981. Teratocarcinomas and other neoplasms as developmental defects in gene expression. *Adv. Cancer Res.* **34**: 211.

Palmiter, R.D., R.L. Brinster, R.E. Hammer, M.E. Trumbauer, M.G. Rosenfeld, N.C. Birnberg, and R.M. Evans. 1982. Dramatic growth of mice that develop from eggs microinjected with metallothionein-growth hormone fusion genes. *Nature* **300**: 611.

Palmiter, R.D., G. Norstedt, R.E. Gelinas, R.E. Hammer, and R.L. Brinster. 1983. Metallothionein human GH fusion genes stimulated growth of mice. *Science* **222**: 809.

Percy, D.H. and A.M. Jonas. 1971. Incidence of spontaneous tumors in CD$^{(R)}$-1 HaM/ICR mice. *J. Natl. Cancer Inst.* **16**: 1045.

Ringold, G.M. 1983. Regulation of mouse mammary tumor virus gene regulation by glucocorticoid hormones. *Curr. Top. Microbiol. Immunol.* **106**: 79.

Stewart, T.A., P.K. Pattengale, and P. Leder. 1984. Spontaneous mammary adenocarcinomas in transgenic mice that carry and express fusion genes. *Cell* **38**: 627.

Stewart, T.A., E.F. Wagner, and B. Mintz. 1982. Human beta globin sequences injected into mouse eggs, retained in adults and transmitted to progeny. *Science* **217**: 1046.

Wagner, E.F., T.A. Stewart, and B. Mintz. 1981. The human beta globin gene and a functional viral thymidine kinase gene in developing mice. *Proc. Natl. Acad. Sci. U.S.A.* **78**: 5016.

Developmental Synergism: The Creation of Novel Tissue Specificities Via the Expression of Growth Hormone Genes in Transgenic Mice

RONALD M. EVANS,* LARRY W. SWANSON,* AND MICHAEL G. ROSENFELD†
*Salk Institute for Biological Studies
La Jolla, California 92037
†University of California, San Diego, School of Medicine
La Jolla, California 92093

OVERVIEW

The molecular basis for neuron-specific gene expression has been examined by analyzing the expression of metallothionein I growth hormone fusion gene in transgenic animals. Based on the unexpected differential expression of metallothionein I in astrocytes and of the fusion gene in specific neurons, irrespective of chromosomal insertion site, we propose that regulatory elements in both parts (5′ and 3′ GH sequences) of the fusion gene interact to generate the observed pattern of gene expression, a phenomenon referred to as developmental synergism.

INTRODUCTION

Gene transfer studies have demonstrated the existence of nucleotide sequences necessary for eukaryotic gene transcription (Brethnach and Chambon 1981; Grosveld et al. 1982; McKnight and Kingsbury 1982; Dierks et al. 1983), and specific regulatory sequences have also been identified in several inducible genes of higher eukaryotics by in vitro mutagenesis and gene transfer analysis (Pelham 1982; Raag and Weisman 1983; Karin et al. 1984). Furthermore, the transfer of a variety of intact and fusion genes into differentiated cells that express related endogenous genes has demonstrated the potential of enhancer sequences to direct tissue and cell-specific expression (Renkawitz et al. 1982; Spandidos and Paul 1982; Chao et al. 1983; Gillies et al. 1983; Oi et al. 1983; Queen and Baltimore 1983; Walker et al. 1983). However, this approach is limited by the availability of appropriate cultured cell types, and to genes that do not require sequential activation by factors produced in earlier lineages of the recipient cell.

The recent development of methods for introducing foreign genes into the germline of mice (Gordon et al. 1980; Brinster et al. 1981; Costantini and Lacy 1981) and *Drosophila* (Spradling and Rubin 1982) provides a complementary approach for studying mechanisms underlying inducible and developmental gene regulation. DNA injected into the fertilized egg can become covalently integrated into the host genome at apparently random sites and can be passed from one generation to the next in a Mendelian way. In some instances, these transferred genes are transcribed into mRNA, which in turn directs the synthesis of an appropriate protein. Transgenic animals expressing foreign genes can be used as a tool to test models of the role played by specific DNA sequences in determining tissue-specific expression.

The expression of a gene may be controlled at many levels, and thus the achievement of appropriate expression may not localize the sequences that determine such specificity. It might therefore be possible to evaluate the potential function, in tissue-specific expression, of sequences within a gene by fusing it with a heterologous promoter. We have utilized such an approach to show there are sequences in the rat growth hormone gene that are capable of influencing tissue-specific expression.

RESULTS

We have previously described the introduction of rat growth hormone (rGH) fusion genes into fertilized mouse eggs, which are then implanted into the oviducts of pseudopregnant mice (Palmiter et al. 1982). In a significant proportion of the eggs (20–30%), the fusion gene is stably and heritably incorporated into the genome. Most of the animals (approximately 80%) that developed from these eggs synthesized rat GH mRNA. Furthermore, many of these mice contain high levels of rGH in the blood, and may grow to almost twice the size of littermate controls. In these experiments, the structural part of the GH gene was fused to the promoter region of the mouse metallothionein I (MT-I) gene (Fig. 1A). Recently, similar experiments utilizing metallothionein human GH gene fusions have been reported (Palmiter et al. 1983). In contrast, introduction of the intact rat GH gene, containing 3 kb of 5′ flanking sequences does not produce gigantism and thus far is not expressed in any tissue analyzed. In the adult mouse, the MT-I gene has a characteristic pattern of expression in various tissues (Danielson et al. 1982a,b). Although many tissues are capable of producing metallothionein mRNA, the liver is the primary site of synthesis (Durnam and Palmiter 1981). Tissues such as the brain express the MT-I gene at levels 40-fold (or more) lower than the liver (Fig. 1B). The growth hormone gene in adult mammals appears to be expressed entirely within somatotrophic cells, which constitute approximately 40% of the anterior pituitary. It was therefore unexpected to find exceptionally high levels of the fusion gene expressed in the brain of transgenic animals (Fig. 1C). These findings led us to examine in more detail the cellular distribution of fusion gene expression in transgenic animals and to determine if this specificity is a consequence of sequences in the gene.

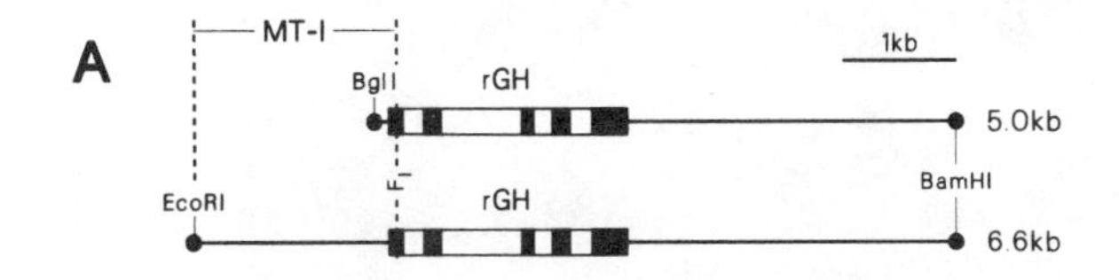

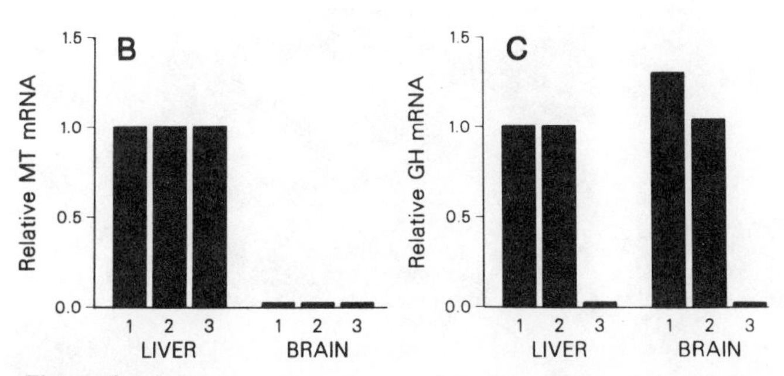

Figure 1
Structure and expression of metallothionein-growth hormone fusion genes. (*A*) Restriction fragments from MTGH fusion genes that were injected into the male pronucleus of fertilized eggs. Relative amount of RNA in livers and brains of transgenic (*1* and *2*) and normal control (*3*) mice as detected by slot blot hybridization with nick translated MT-I cDNA (*B*) or rat growth hormone cDNA as probe (*C*).

Expression in the Brain

Based upon the pattern of coexpression of MT and GH gene products, histochemical analysis revealed that the fusion gene gave a highly restricted pattern of expression in several anatomically defined areas in the brain. The most obvious pattern of GH cell staining in transgenic mice was observed in the hypothalamus. Because the peptide content and neuroendocrine function of cells in this part of the brain are particularly well understood (Swanson and Sawchenko 1983), attention was focused initially on the precise cellular distribution of hypothalamic GH immunostaining. A detailed examination of eight animals representing offspring from five generations of a single transgenic parent (MGH-10) showed an identical distribution of GH-stained neuronal cell bodies in seven of the animals. GH-stained cells were concentrated in three parts of the hypothalamus. The most obvious group of cells consisted of large neurons in the paraventricular (PVH) and supraoptic (SO) nuclei (Fig. 2A,B), which synthesize oxytoxin and vasopressin, project to the posterior pituitary, and are involved in the integration of neuroendocrine and autonomic responses (Swanson and Sawchenko 1983). In all cases, neuronal staining was blocked by the addition of rGH to both of the anti-GH sera used. No glial

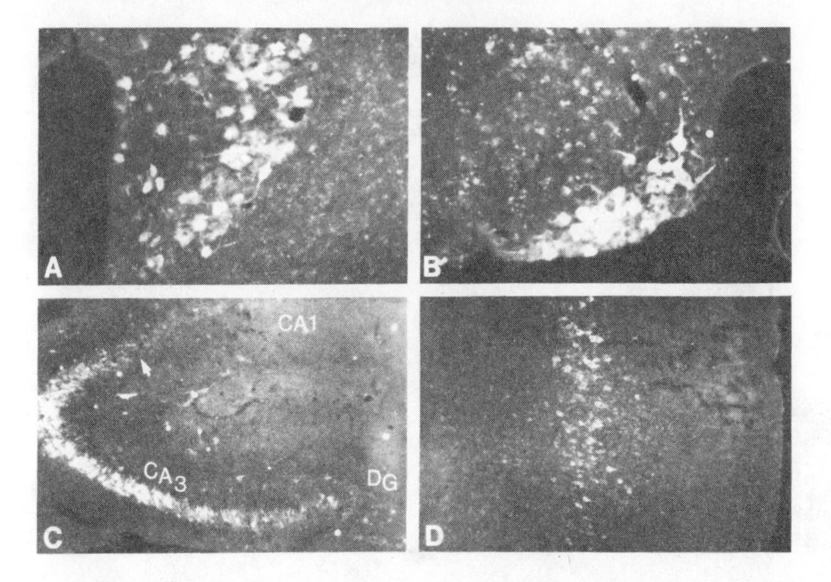

Figure 2
Frontal sections through the hypothalamus of an MGH-10 animal. Large neurons in the paraventicular (*A*) and supraoptic (*B*) nuclei stained with an antiserum to mouse growth hormone (x150). (*C*) Immunostaining in cell bodies in the hippocampus are confined to the pyramidal cell layer. The arrowhead shows the boundary between fields CA1 and CA2 (x75). (*D*) Stained pyramidal cell bodies in layer V of the junctional region between the somatic sensory (upper half) and motor (lower half) cortex on the left side of the brain.

or vascular cell types were stained. These results demonstrate that GH is consistantly expressed in specific subpopulations of neurons in the brain.

In accord with the prominent but highly restricted expression of the GH fusion gene in the hypothalamus, specific expression was observed in several areas outside the hypothalamus, including regions of the neocortex, hippocampus, and olfactory cortex in the MGH-10 pedigree. All animals examined showed a similar pattern of GH immunostaining that was greatly enhanced by colchicine pretreatment, which blocks axonal transport. Immunoactive pyramidal cells restricted to layer V of the neocortex (Fig. 2D) were found in all animals expressing the fusion gene in liver and hypothalamus. This cortex is broadly divided into six layers with unique cell types that are morphologically and physiologically distinct. Bright immunostaining in this region is highly localized and does not appear to enter other layers.

The hypothalamus and neocortex are extensively interconnected with a ring of forebrain structures called the limbic region, which is concerned with the maintenance of homeostasis. One part of this region is the hippocampal formation, which is composed of the hippocampus proper, the dentate gyrus, the subicular region, and the entorhinal area. In the brains of transgenic mice, specific immunostaining is also seen in the pyramidal cells of hippocampal field CA_3 (Fig. 2C), which is a region that is dominated by mossy fiber afferents from the adjacent dentate gyrus. Although pyramidal cells in the other hippocampal fields (CA1, 2, and 4) were negative, some interneurons were stained here, as well as in the hilus of the dentate gyrus (Fig. 2C).

To determine whether the site of chromosomal integration of the fusion gene is critical in determining the pattern of MGH expression in neurons, six other integrations of the MTrGH gene, and two integrations of the MThGH gene, were examined in a total of eleven animals. In each brain, the pattern of GH immunostaining in the hypothalamus was examined in detail and was identical to that described for the MGH-10 pedigree, although the intensity of staining was variable. For example, integration of the fusion gene into the dwarf little (*Lit*) mouse, which has an inherited mutation in expression of the endogenous GH gene[25], resulted in weak immunostaining in two mice from a pedigree, and no detectable staining in a third animal. On the other hand, immunostaining in two different MThGH integrations was more intense than in those expressing the rGH fusion gene.

In summary, 16 of the 18 transgenic mice analyzed, representing nine different integrations of the rGH and hGH fusion genes, showed the same cellular distribution of GH in the brain (Fig. 3). An identical pattern of GH immunostaining was seen in males and females, as well as in albino, agouti, and black strains of mice. These data strongly suggest that the precise site of integration into the genome may not play a critical role in determining which neurons in the brain express the fusion gene, but may well modulate the absolute level of gene expression in these cells.

Metallothionein Distribution in Brain

If sequences in the metallothionein (MT) promoter determine the specific pattern of expression of the MGH fusion gene in the brain of transgenic mice, then one might predict that the endogenous MT-I gene would be expressed with an identical distribution. An antiserum to rat liver, MT-I was used to identify the pattern of MT-I immunostaining in normal and transgenic mice. This antiserum consistently stained small cells with the appearance of astrocytes in all parts of the white and gray matter, both in normal and transgenic animals. This tentative identification was confirmed in double immunostaining experiments with a monoclonal antibody (Haan et al. 1982) to the astrocyte marker, S-100, and a polyclonal antiserum (raised in rabbit) to MT-I. Essentially all cells were doubly labeled in this material (data not shown), and no evidence for MT staining of neurons was obtained. These results indicate that although the fusion gene and MT-I appear to be co-expressed in hepatocytes, their expression is independently and differently regulated in the brain.

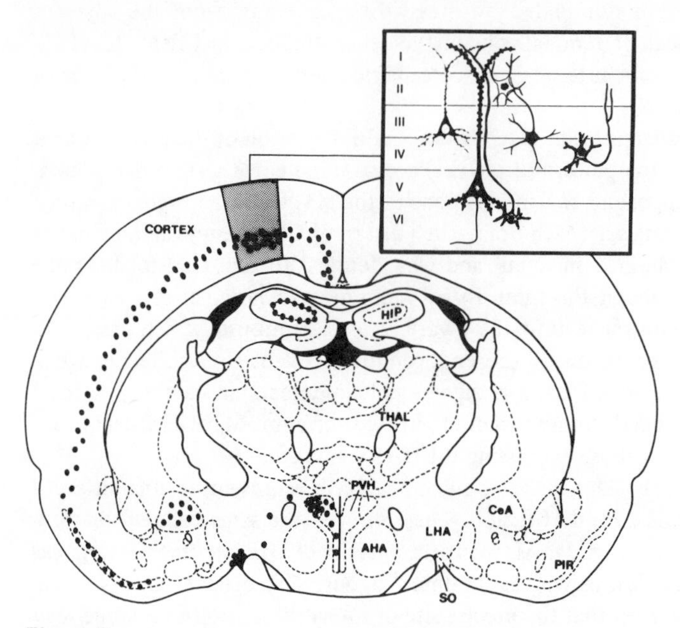

Figure 3
This diagram summarizes the discrete neuronal localization (dots) of GH immunostaining in a frontal section through the forebrain of a rodent. The insert shows that while the cerebral cortex contains many cell types that are distributed through six layers, GH staining is confined to large pyramidal cells (dark) in layer V. (AHA) Anterior hypothalamic area; (CeA) central nucleus of the amygdala; (HIP) hippocampus; (LHA) lateral hypothalamic area; (PIR) piriform cortex; (PVH) paraventicular nucleus; (SO) supraoptic nucleus; (THAL) thalamus.

DISCUSSION

One approach to evaluate the role of sequences in determining developmental specificity is to reintroduce intact normal genes into animals and to determine whether their expression is identical to that of the endogenous gene. Because tissues generally contain a number of specialized cell types, determination of developmental specificity should be examined at the cellular level.

To examine the potential functions of GH sequences in dictating developmental expression of the GH gene, the promoter of the MT-I gene, which is widely expressed in visceral tissue, was fused to the structural portion of the GH gene and 3 kb of downstream flanking chromosomal sequences.

Unexpectedly, whereas MT is localized to glial cells (astrocytes) in the brain, expression of the MGH fusion gene is confined to neurons. Furthermore, detectable neuronal expression is restricted to several highly localized regions that include pyramidal cells in layer V of the neocortex and field CA3 of the hippocampus, magnocellular neurosecretory cells of the hypothalamus, the suprachiasmatic nucleus, the central nucleus of the amygdala, the compact zone of the substantia nigra and adjacent parts of the ventral tegmental area, and the dorsal motor nucleus of the vagus (Fig. 3). Because these areas are not known to share common developmental influences, do not appear to use a common neurotransmitter, and do not manifest functional or cytological similarities, there is at present no obvious explanation for why the fusion gene is coordinately expressed in these particular neurons. Because metallothionein expression in the brain fails to colocalize with expression of the fusion gene, it is tempting to propose a model that requires sequences within both the 5′ flanking and 3′ structural region of the fusion gene for the observed cellular specificity. According to this hypothesis, sequences present in the growth hormone chromosomal fragment are capable of promoting expression outside of the anterior pituitary, which appears to be the only location for endogenous growth hormone gene expression. This would further suggest that specific developmental expression of the growth hormone gene might require interaction between 5′ flanking chromosomal sequences and those present in downstream regions of the gene. Chromosomal position of the fusion genes as a means to control growth hormone specificity is essentially excluded because of the identical pattern of expression of the fusion gene in the brains of unrelated transgenic animals.

It is also possible that the effect of sequences in the fusion gene might be modulated further by chromosomal position since the absolute level of GH expression appears to vary in different integrations. Whatever the explanation, the results presented here suggest a model in which combinations of sequence components might give rise to observed developmental and tissue specificities (Fig. 4). According to

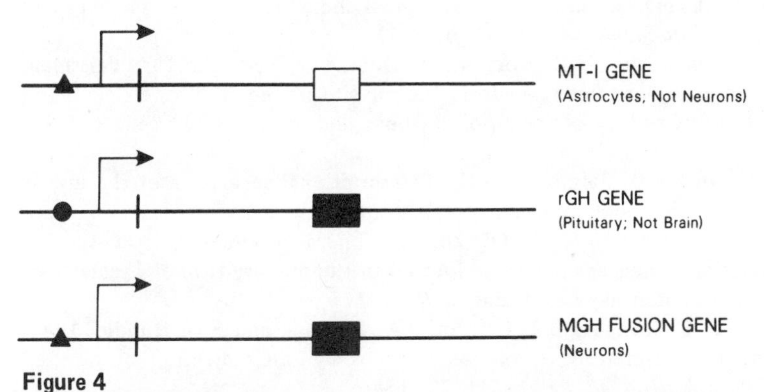

Figure 4
Model of developmental synergism of GH and MT chromosomal regions. Symbols (▲, ●, ▲) represent putative genetic elements capable of influencing gene expression. Horizontal lines schematically represent genes, vertical lines represent sites of gene fusion, and arrows represent transcription initiation sites.

this model, the determination of the developmental specificity in some genes (such as rGH) may be due to the effects of several distinct regulatory elements. The separation of these elements may alter or abolish specificity whereas their linkage to elements from other genes may generate novel specificities such as that observed with the MGH gene. If this is true, it may be possible to introduce specific gene products into the brains of transgenic mice, providing a potential means to manipulate genetically CNS function.

It currently is not possible to determine whether the observed pattern of MGH gene expression in the nervous system is novel or corresponds to as yet undescribed endogenous gene products. Nonetheless, the acquisition of a specificity distinct from either parent gene suggests another level at which the genome may expand its developmental diversity.

REFERENCES

Brethnach, R. and P.A. Chambon. 1981. Organization and expression of eukaryotic split genes coding for proteins. *Annu. Rev. Biochem.* **50**: 349.

Brinster, R.L., H.Y. Chen, M. Trumbauer, A.W. Senear, R. Warren, and R.D. Palmiter. 1981. Somatic expression of herpes thymidine kinase in mice following injection of a fusion gene into eggs. *Cell* **27**: 223.

Chao, M.V., P. Mellon, P. Charnay, T. Maniatis, and R. Axel. 1983. The regulated expression of β-globin genes introduced into mouse erythroleukemid cells. *Cell* **32**: 483.

Costantini, F. and E. Lacy. 1981. Introduction of a rabbit β-globin gene into the mouse germ line. *Nature* **294**: 92.

Danielson, K.G., S. Ohi, P.C. Huang. 1982a. Immunochemical detection of metallothionein in specific epithelial cells of rat organs. *Proc. Natl. Acad. Sci. U.S.A.* **79**: 2301.

———. 1982b. Immunochemical localization of metallothionein in rat liver and kidney. *J. Histochem. Cytochem.* **30**: 1033.

Dierks, P., A. van Ooyen, M.D. Cochran, C. Dobkin, J. Reiser, and C. Weissman. 1983. Three regions upstream from the cap site are required for efficient and accurate transcription of the rabbit β-globin gene in mouse 3T6 cells. *Cell* **32**: 692.

Durnam, D.M. and R.D. Palmiter. 1981. Transcriptional regulation of the mouse metallothionein I gene by heavy metals. *J. Biol. Chem.* **256**: 5712.

Gillies, S.D., S.L. Morrison, V.T. Oi, and S. Tonegawa. 1983. A tissue-specific transcription enhancer element is located in the major intron of a rearranged immunoglobulin heavy chain gene. *Cell* **33**: 717.

Gordon, J.W., G.A. Scangos, D.J. Plotkin, J.A. Barbosa, and F.H. Ruddle. 1980. Genetic transformation of mouse embryos by microinjection of purified DNA. *Proc. Natl. Acad. Sci. U.S.A.* **77**: 7380.

Grosveld, G.C., A. Rosenthal, and R.A. Flavell. 1982. Sequence requirements for the transcription of the rabbit β-globin gene in vivo: The −80 region. *Nucleic Acids Res.* **10**: 4951.

Haan, E.A., B.D. Boss, and W.M. Lowan. 1982. Production and characterization of monoclonal antibodies against the "brain-specific" proteins 14-3-2 and S-100. *Proc. Natl. Acad. Sci. U.S.A.* **79**: 7585.

Karin, M., A. Haslinger, H. Holtgreve, R. Richards, P. Krauter, H. Westphal, and M. Beato. 1984. Characterization of DNA sequences through which cadmium and glucocorticoid hormones induce human metallothionein II gene. *Nature* **308**: 513.

McKnight, S.L. and R. Kingsbury. 1982. Transcriptional control signals of a eukaryotic protein-coding gene. *Science* **217**: 316.

Oi, V., S.L. Morrison, L.A. Herzenberg, and P. Berg. 1983. Immunoglobulin gene expression in transformed lymphoid cells. *Proc. Natl. Acad. Sci. U.S.A.* **80**: 825.

Palmiter, R.D., G. Norstedt, R.E. Gelinas, R.E. Hammer, and R.L. Brinster. 1983. Metallothionein human GH fusion genes stimulate growth of mice. *Science* **222**: 809.

Palmiter, R.D., R.L. Brinster, R.E. Hammer, M.E. Trumbauer, M.G. Rosenfeld, N.C. Birnberg, and R.M. Evans. 1982. Dramatic growth of mice that develop from eggs microinjected with metallothionein growth hormone fusion genes. *Nature* **300**: 611.

Pelham, H. 1982. A regulatory upstream promoter element in the drosophila Hsp 70 heat-shock gene. *Cell* **30**: 517.

Queen, C. and D. Baltimore. 1983. Immunoglobulin gene transcription is activated by downstream sequence elements. *Cell* **33**: 741.

Raag, H. and C. Weisman. 1983. Not more than 117 base pairs of 5′-flanking sequence are required for inducible expression of a human 1FN- gene. *Nature* **303**: 439.

Renkawitz, R., H. Beug, T. Graf, P. Matthais, M. Greg, and G. Schutz. 1982. Expression of a chicken lysozyme recombinant gene is regulated by progesterone and dexamethasone after microinjection into oviduct cells. *Cell* **31**: 167.

Spandidos, D.A. and J. Paul. 1982. Transfer of human globin genes to eythioleukemia mouse cells. *Eur. Mol. Biol. Organ. J.* **1**: 15.

Spradling, A.C. and G.M. Rubin. 1982. Transposition of clonal P elements into Drosophila germ line chromosomes. *Science* **218**: 341.

Swanson, L.W. and P.E. Sawchenko. 1983. Hypothalamic integration: Organization of the paraventicular and supraoptic nuclei. *Annu. Rev. Neurosci.* **6**: 269.

Walker, M., T. Edlund, A. Boulet, and W. Rutten. 1983. Cell specific expression controlled by the 5′ flanking region of insulin and chymotrypsin genes. *Nature* **306**: 557.

Homeo-box Sequences — Relevance to Vertebrate Developmental Mechanisms

FRANK H. RUDDLE,* CHARLES P. HART,* AND WILLIAM McGINNIS†
*Department of Biology
†Department of Molecular Biophysics and Biochemistry
Yale University
New Haven, Connecticut 06511

INTRODUCTION

Reverse genetics provide a means for studying complex regulatory mechanisms such as those involved in development (Ruddle 1984). The basic idea is to identify genetic mutants that affect development processes, map the mutant, and then isolate the gene of interest using molecular cloning techniques. The gene's functional attributes can then be ascertained by nucleotide sequence analysis, coupled with the deduction of the protein structure of the coded product, the ontogenetic expression of the mRNA at the cellular and tissue level by in situ hybridization, and an immunological characterization of the specified protein product. Nowhere has this approach proven more successful than in the case of the homeotic genes of *Drosophila* (Lewis 1978; Bender et al. 1983). The homeotic genes have been shown to control the developmental determination of groups of cells with respect to their morphogenetic fates. Many of the genes are clustered in two gene complexes, the Antennapedia-complex, (ANT-C) and the Bithorax-complex (BX-C). Genes within the two complexes include *Antennapedia* (*Antp*), *Ultrabithorax* (*Ubx*), and *fushi-tarazu* (*ftz*). The *ftz* gene regulates the number of embryonic segments, whereas *Antp* and *Ubx* control the morphogenesis of specific body segments (Lewis 1978; Kaufman et al. 1980; Wakimoto and Kaufman 1984).

Recently, it has been shown that a characteristic 180 base pair sequence, named the homeo-box, is present within homeotic genes and gene complexes (McGinnis et al. 1984a). The homeo-box has the following set of properties: Its sequence is highly conserved within *Drosophila melanogaster* at the 70–90% level. There are more than seven copies and those so far characterized are associated with the homeotic gene set. In all instances studied, the nucleotide sequence is part of an open reading frame and the sequence has been detected in mRNA by sequence homology. Temporal and spatial expression of the mRNA species correlate well with the known morphogenetic properties of the individual homeotic genes. A subdomain of the homeo-box sequence has the general properties of DNA binding proteins (see below).

Homeo-box sequences are also conserved in a broad range of invertebrate and vertebrate species (Carrasco et al. 1984; McGinnis et al. 1984b). McGinnis and coworkers have shown by DNA hybridization studies that highly conserved homeo-box sequences exist in other insects, crustaceans, annelids, amphibians, birds, and mammals, including man and the laboratory mouse (McGinnis et al. 1984b). Moreover, animals with a nonmetameric body plan, such as the nonsegmented nematodes, show only weak homology to homeo-box sequences, if at all. This high level of correlation between the occurrence of homeo-box sequences and metamerism suggests that the homeo-box sequence may play a significant role in the pattern formation of segmentation and the developmental determination of the segments themselves. This assumption is further strengthened by the fact that some 10–15 copies of sequences are regularly detected in the metameric species tested.

The conservation of the homeo-box sequences in the metameric forms of the proto- and deutero-stomia suggests the potential for a comparative genetic approach to the analysis of developmental mechanisms in higher organisms. If gene regulation mechanisms pertaining to development have had their origins in primitive metazoan forms, it is likely that they have been preserved in derivative forms as well. This is likely because they involve the concerted interaction of numerous genes in time and space throughout the ontogenetic process. It is unlikely that these complex interactions once established could be easily changed. Thus according to this view, the analysis of developmental systems in lower forms where such analysis can be efficiently pursued, as for example by a reverse genetic approach, could then be extended to higher forms through the application of a comparative genetics approach.

The discovery of homeo-box conserved species in the mammals provides a means of testing their general biological role. The genetics of man and the laboratory mouse are highly developed, and in addition, *Mus musculus* provides an ideal system for developmental studies at the experimental level. In this paper, we shall review the data that has been recently obtained on homeo-box sequences in these two species and project the likely course of future investigations designed to explain the developmental relevance of the homeo-box sequences.

MOUSE HOMEO-BOX GENES

Mouse homeo-box sequences were detected by Southern blotting hybridization analysis using homeo-box probes isolated from the *Drosophila* Antennapedia and Bithorax complexes. Mouse homeo-box sequences were isolated by screening a λ Charon 28 mouse genomic library with *Drosophila* homeo-box probes from Antennapedia and Ultrabithorax either singly or in combination. Twelve mouse sequences were isolated of which to date at least seven have proven to be unique.

One mouse homeo-box sequence, termed Mo-10 has been sequenced and the nucleotide and the deduced peptide sequences are presented in Figures 1 and 2, respectively (McGinnis et al. 1984c). The degree of conservation of nucleotide sequence is considerable having a value of 75% and containing an open reading frame. The deduced peptide homology is even greater at 85%. It is also evident that where

```
                      -21                        -1
Antp                                                          ATT TAC TTG GAA CCA ACA GAA
Mo -10                                                        GGA CCG CCT GGG CAG GCC TCG
      -56
Hu-1    GAA TAA GTG TCG TTG CGG CTT TCC TCT ATC TGC TCC AGA TAT GAC GGG CCG GAC
Hu-2    TGA CCG CAG GCC TCA GCA TCT CCA CTC TGC GTA ACA GGT TCC TCC TTT GGG CCC
AC1                                                           GGC GTG GGC TAC GGG TCG GAC
MM3                                                           CTT TCT CTT GCA GGG GCG GAC

         1                                   30                                          60
Antp    CGC AAA CGC GGA AGG CAG ACA TAC ACC CGG TAC CAG ACT CTA GAG CTA GAG AAG GAG TTT
Mo-10   TCC AAG CGC GGC CGC ACG GCG TAC ACG AGG CCG CAG CTG GTA GAG CTG GAG AAG GAG TTC
Hu-1    GGG AAA AGG GCC CGG ACC GCG TAT ACC CGC TAC CAG ACC CTG GAG CTG GAA AAG GAG TTC
Hu-2    ACG GCC GGA GGA CGC CAG ACA TAC ACA CGT TAC CAG ACG CTG GAG CTG GAG AAG GAG TTT
AC1     AGG AGG AGA GGA CGC CAG ATC TAT TCC CGT TAC CAA ACT CTG GAG CTC GAG AAA GAA TTT
MM3     AGG AAG AGG GGT CGC CAG ACC TAC ACA AGG TAC CAG ACC CTG GAG CTG GAG AAG GAG TTT

        61                                   90                                         120
Antp    CAC TTC AAT CGC TAC TTG ACC CGT CGG CGA AGG ATC GAG ATC GCC CAC GCC CTG TGC CTC
Mo-10   CAC TTC AAC CGC TAC CTA ATG CGG CCG CGC CGG GTG GAG ATG GCC AAC CTG CTG AAC CTC
Hu-1    CAC TTC AAC CGC TAC CTG ACC CGG CGA CGG CGC ATC GAG ATC GCC CAC GCA CTC TGC CTG
Hu-2    CAC TAC AAT CGC TAC CTG ACG CGG CGG CGG CGC ATC GAG ATC GCG CAC GCC CTG TGC CTG
AC1     CAC TTC AAT CGC TAC CTG ACC CGG CGC AGG AGG ATC GAG ATC GCC AAT GCA CTT TGT CTC
MM3     CAC TTT AAC CGC TAC CTG ACC CGG CGG AGG CGC ATC GAG ATC GCC CAC GTT CTG TGT CTG

       121                                  150                                         180
Antp    ACG GAG CGC CAG ATA AAG ATT TGG TTC CAG AAT CGG CGC ATG AAG TGG AAG AAG GAG AAC
Mo-10   ACC GAG CGC CAG ATC AAG ATC TGG TTT CAG AAC CGG CGC ATG AAG TAC AAG AAA GAC CAG
Hu-1    TCC GAG CGC CAG ATC AAG ATC TGG TTC CAG AAC CGG CGC ATG AAG TGG AAG AAG GAC AAC
Hu-2    ACG GAG AGG CAG ATC AAG ATA TGG TTC CAG AAC CGA CGC ATG AAG TGG AAA AAG GAG AGC
AC1     ACA GAG CGG CAG ATC AAA ATC TGG TTC CAG AAC AGG AGG ATG AAA TGG AAA AAG GAG AGC
MM3     ACC GAG CGA CAA ATC AAA ATC TGG TTC CAG AAC CGC AGG ATG AAA TGG AAA AAG GAA AAC

       181                     201
Antp    AAG ACG AAG GGC GAG CCG GAT
Mo-10   AAG GGC AAA GGC ATG CTG ACC                                             240
Hu-1    AAA TTG AAA AGT ATG AGC CTG GCT ACA GCT GGC AGC GCT TCC AGC CCT GAG CCC GCC CAG AGG AGC CAA
Hu-2    AAA CTG CTC AGC GCG TCT CAG CTC AGT GCC GAG GAG GAG GAA GAA AAA CAG GCC GAG TGA AGG TGC TGG
AC1     AAC CTC TCA TCT ACC CTT
MM3     AAG GCA TCC AGC CCT TCC TCT AAC AGC CAG GAA AAG CAG GAG ACT GAG GAA GAG GAG GAG GAA TGA AGT
```

Figure 1
Homeo-box nucleic acid sequences of Antennapedia (*Antp*) from *Drosophila*, Mo-10 from mouse, Hu-1 and -2 from human, and *CC1* and *MM3* from *Xenopus*. The homeo-box domain extends between nucleotides 1 and 180. See text for relevant references.

												−7						−1
Antp												Ile	Tyr	Leu	Glu	Pro	Thr	Glu
Mo-10												Gly	Pro	Pro	Gly	Gln	Ala	Ser
Hu-1	Glu	Trm	Val	Ser	Leu	Arg	Leu	Ser	Ser	Ile	Cys	Ser	Arg	Tyr	Asp	Gly	Pro	Asp
Hu-2	Trm	Pro	Glm	Ala	Ser	Ala	Ser	Pro	Leu	Cys	Val	Thr	Gly	Ser	Ser	Phe	Gly	Pro
AC1												Gly	Val	Gly	Tyr	Gly	Ser	Asp
MM3												Leu	Ser	Leu	Ala	Gly	Ala	Asp

	1									10										20
Antp	Arg	Lys	Arg	Gly	Arg	Gln	Thr	Tyr	Thr	Arg	Tyr	Gln	Thr	Leu	Glu	Leu	Glu	Lys	Glu	Phe
Mo-10	Ser	Lys	Arg	Gly	Arg	Thr	Ala	Tyr	Thr	Arg	Pro	Gln	Leu	Val	Glu	Leu	Glu	Lys	Glu	Phe
Hu-1	Gly	Lys	Arg	Ala	Arg	Thr	Ala	Tyr	Thr	Arg	Tyr	Gln	Thr	Leu	Glu	Leu	Glu	Lys	Glu	Phe
Hu-2	Thr	Ala	Gly	Gly	Arg	Gln	Thr	Tyr	Thr	Arg	Tyr	Gln	Thr	Leu	Glu	Leu	Glu	Lys	Glu	Phe
AC1	Arg	Arg	Arg	Gly	Arg	Gln	Ile	Tyr	Ser	Arg	Tyr	Gln	Thr	Leu	Glu	Leu	Glu	Lys	Glu	Phe
MM3	Arg	Lys	Arg	Gly	Arg	Gln	Thr	Tyr	Thr	Arg	Tyr	Gln	Thr	Leu	Glu	Leu	Glu	Lys	Glu	Phe
														hpo	Ala			hpo *	*	hpo *

	21									30	——	——	——	Helix 2	——	——	——	——		40
Antp	His	Phe	Asn	Arg	Tyr	Leu	Thr	Arg	Arg	Arg	Arg	Ile	Glu	Ile	Ala	His	Ala	Leu	Cys	Leu
Mo-10	His	Phe	Asn	Arg	Tyr	Leu	Met	Arg	Pro	Arg	Arg	Val	Glu	Met	Ala	Asn	Leu	Leu	Asn	Leu
Hu-1	His	Phe	Asn	Arg	Tyr	Leu	Thr	Arg	Arg	Arg	Arg	Ile	Glu	Ile	Ala	His	Ala	Leu	Cys	Leu
Hu-2	His	Tyr	Asn	Arg	Tyr	Leu	Thr	Arg	Arg	Arg	Arg	Ile	Glu	Ile	Ala	His	Ala	Leu	Cys	Leu
AC1	His	Phe	Asn	Arg	Tyr	Leu	Thr	Arg	Arg	Arg	Arg	Ile	Glu	Ile	Ala	Asn	Ala	Leu	Cys	Leu
MM3	His	Phe	Asn	Arg	Tyr	Leu	Thr	Arg	Arg	Arg	Arg	Ile	Glu	Ile	Ala	His	Val	Leu	Cys	Leu

					Ile/Val *			hpo *												
	41	——	——	——	Helix 3	——	——	——		50										60
Antp	Thr	Glu	Arg	Gln	Ile	Lys	Ile	Trp	Phe	Gln	Asn	Arg	Arg	Met	Lys	Trp	Lys	Lys	Glu	Asn
Mo-10	Thr	Glu	Arg	Gln	Ile	Lys	Ile	Trp	Phe	Gln	Asn	Arg	Arg	Met	Lys	Tyr	Lys	Lys	Asp	Gln
Hu-1	Ser	Glu	Arg	Gln	Ile	Lys	Ile	Trp	Phe	Gln	Asn	Arg	Arg	Met	Lys	Trp	Lys	Lys	Asp	Asn
Hu-2	Thr	Glu	Arg	Gln	Ile	Lys	Ile	Trp	Phe	Gln	Asn	Arg	Arg	Met	Lys	Trp	Lys	Lys	Glu	Ser
AC1	Thr	Glu	Arg	Gln	Ile	Lys	Ile	Trp	Phe	Gln	Asn	Arg	Arg	Met	Lys	Trp	Lys	Lys	Glu	Arg
MM3	Thr	Glu	Arg	Gln	Ile	Lys	Ile	Trp	Phe	Gln	Asn	Arg	Arg	Met	Lys	Trp	Lys	Lys	Glu	Asn

	61						67																
Antp	Lys	Gly	Lys	Gly	Met	Leu	Thr																
Mo-10	Lys	Gly	Lys	Gly	Met	Leu	Thr																
Hu-1	Lys	Leu	Lys	Ser	Met	Ser	Leu	Ala	Thr	Ala	Gly	Ser	Ala	Ser	Ser	Pro	Glu	Pro	Ala	Gln	Arg	Ser	Pro
Hu-2	Lys	Leu	Leu	Ser	Ala	Ser	Gln	Leu	Ser	Ala	Glu	Glu	Glu	Glu	Glu	Lys	Gln	Ala	Glu	Trm	Arg	Cys	Trp
AC1	Asn	Leu	Ser	Ser	Thr	Leu																	
MM3	Lys	Ala	Ser	Ser	Pro	Ser	Ser	Asn	Ser	Gln	Glu	Lys	Gln	Glu	Thr	Glu	Glu	Glu	Glu	Glu	Glu	Stop	

Figure 2

Comparison of the amino acid sequence of the murine homeo-box sequence Mo-10, human homeo-box sequences Hu-1 and Hu-2, and *Xenopus* homeo-box sequences AC1 and MM3 with the *Drosophila* Antennapedia homeo-box. The sequences are deduced from nucleotide base sequences. The sequence 40–57 shows homology to the yeast *MAT* loci. The sequence 31–49 shows a similarity to the DNA binding domains of various prokaryotic regulatory proteins. Domains that correspond to helical domains in the prokaryotic repressor proteins are designated helix 2 and helix 3 following the λ Cro convention. Hpo identifies residues that are generally hydrophobic in the prokaryotic repressors. Ala and Ile/Val are invariably found in the prokaryotic repressors at codon sites corresponding to 35 and 45, respectively. In the prokaryotic regulatory proteins, the sites corresponding to those starred in the above figure show a high level of conservation. They are believed to be important for a helix stability. See text for references.

amino acids differ between *Drosophila* and *Mus*, they are likely to be conserved in terms of their functional properties. We can expect that the other murine homeo-box sequences will also show a high degree of conservation based on hybridization homology.

In addition to conservation of sequence homology within the homeo-box domains, evidence is accumulating suggesting homology in flanking genomic DNA sequences between species for corresponding homeotic genes. Using flanking genomic sequences as a probe in Southern hybridization experiments against restricted mouse, human, and Chinese hamster DNA, one observes hybridizing fragments in all three species. These kinds of experiments suggest that for a particular homeo-box sequence, the flanking sequences are conserved in other species in corresponding or homologous homeotic genes. If cognate homeo-box domains can be demonstrated between mammalian species that again underscores the possibility that they are involved in significant biological functions. Correspondences can also be established by gene mapping. Since M-10 resides in the neighborhood of the *Igk* locus (see below), it will be of interest to determine if its human homolog maps to human chromosome 2, which carries the *Igk* locus. Such correspondences are frequently observed between the human and mouse genomes (Barker et al. 1984; Lee et al. 1984).

Mapping data are useful in establishing or eliminating association with other known genes whose map positions have been determined. More than 50 mutants affecting morphogenesis have been genetically mapped in the mouse. It has been possible to physically map the Mo-10 sequence to a murine chromosomal position. Mapping was established using a panel of mouse x Chinese hamster hybrids that segregate mouse chromosomes. Mo-10 has been mapped to mouse chromosome 6 using a restriction fragment marker unique to the mouse, revealed by hybridization with Mo-10 probe. A subchromosomal assignment could also be made based on the correlation of the Mo-10 marker with subchromosomal fragments of mouse chromosome 6 isolated in mouse x Chinese hamster hybrid cell populations (McGinnis et al. 1984c). These experiments demonstrated that Mo-10 resides proximal to *Igk* in the region of the A and B1 bands. This region of chromosome 6 contains the morphogenetic mutants, hypodactyly (*Hd*) and hydrocephalic-polydactyly (*hpy*) (Green 1981). The *Hd* dominant mutant in heterozygotes results in a reduction of phalanges of the rear feet and shows a more drastic expression in both fore and hind limbs in homozygotes. The *hpy* mutant causes hydrocephally and polydactyly in homozygotes. The distance between *Igk* and the centromere is approximately 30 cM, so that *Hd, hpy,* and Mo-10 may be some considerable distance from one another. Refined linkage mapping can be performed if a restriction site polymorphism were found in association with Mo-10. A search for such a variant is currently underway.

HUMAN HOMEO-BOX GENES

As indicated above, the Mo-10 probe reveals a number of cross hybridizing fragments in restricted human DNA by Southern blot analysis. Between eight and ten human fragments can be routinely detected. Levine and coworkers have recently cloned such fragments using recombinant DNA procedures (Levine et al. 1984). Two homeo-box sequences were found to be in linkage within a distance of 5 kb, and these have been designated Hu-1 and Hu-2. These human homeo-box sequences show remarkable similarities to both mouse and *Drosophila* sequences (see Figs. 1 and 2). The nucleotide sequence, for example, shows an 80% similarity with the Antennapedia homeo-box. Recent unpublished mapping studies suggest that Hu-1 and -2 map to human chromosome 17 (R. V. Lebo, pers. comm.). Comparative genetic mapping studies show no correspondences between mouse chromosome 6 and human chromosome 17. This fact argues against a strict correspondence between mouse homeo-box Mo-10 and human homeo-boxes Hu-1 and -2.

AMPHIBIAN HOMEO-BOX GENES

Homeo-box genes have also been reported to exist in *Xenopus*. In fact, the *Xenopus* gene *AC1* was the first vertebrate homeo-box sequence to be cloned and characterized (Carrasco et al. 1984). Recently, a second *Xenopus* homeo-box gene has been isolated and designated *MM3* (Muller et al. 1984). Both *AC1* and *MM3* possess 180-bp conserved homeo-box sequences showing approximately 80% homology with the Antennapedia homeo-box nucleotide sequence. The predicted amino acid sequences between *AC1*, *MM3*, and *Antp* show even greater homologies at the 90% level (Figure 2). This is again consonant with the argument of conservation serving a biological function. Both *AC1* and *MM3* show a regulated pattern of gene expression during ontogeny as revealed by Northern blotting analysis. *AC1* codes for three transcripts of different molecular size. These are expressed at gastrulation and show different intensities of expression thereafter. *MM3* is most strongly expressed in the oocyte, declines during blastulation, then increases in the gastrula, and continues to be expressed into the tadpole stage. Both 2.1-kb and 1.4-kb molecular forms are expressed in the oocyte.

The data so far obtained for the *Xenopus* homeo-box sequences parallels that reported for man and mouse. A small family of homeo-box sequences exist in the genome, although as yet there is no linkage or mapping data available to suggest their pattern of organization. The nucleotide and amino acid sequences are highly conserved, and there is some suggestion that weak but significant conservation extends outside the 180-bp homeo-box domain per se into the immediate flanking regions. Each homeo-box gene is expressed according to a definite schedule during development, but as yet there is no data reporting on tissue specificity of expression. Altogether, the results for *Xenopus* show striking similarities to the *Drosophila* system, as well as the vertebrates, man and mouse. Most importantly, the expression data provides firm evidence for their genic nature and correlates with specific periods of development.

SIMILARITY OF A HOMEO-BOX SUBDOMAIN TO DNA BINDING DOMAINS

Shepherd et al. (1984) and Laughon and Scott (1984) have reported a nucleotide sequence similarity between *Drosophila* Antennapedia and Ultrabithorax homeo-box sequences and a domain contained within the yeast MAT *a1* and *α2* genes. This is a provocative finding since there is now strong evidence that the MAT *α2* gene product acts as a DNA binding protein which negatively regulates the expression of a set of yeast genes involved in the control of mating type (A. Johnson, pers. comm.). The homeo-box subdomain showing MAT *α2* homology begins at amino acid residue 40 and extends for about 17 residues downstream. The amino acid composition and sequence in part of this subdomain (residues 31-49) also show similarities with a series of prokaryotic DNA binding proteins involved in gene regulation. These include the lac repressor, CAP, and Cro, among others. The subdomain encodes the α-helices thought to interact directly with the DNA. Thus, there is an array of circumstantial evidence suggesting the possibility that the homeo-box product may possess a gene regulatory property mediated by DNA binding. This is an intriguing notion since it reinforces earlier ideas based on genetic theory. These hypotheses stated that homeotic genes performed their morphogenetic control functions by switching on (or off) batteries of genes involved in complex morphogenetic events. This notion is also consistent with recent findings on the spatial and temporal expression of homeotic genes. For example, *fushi tarazu* mutants affect the proper segmentation patterns of larvae. The gene product in terms of cytoplasmic mRNA can be detected with considerable sensitivity by in situ hybridization. Expression is localized to regions of prospective segment furrow formation, and the *ftz* mRNA signal appears, in fact wanes, prior to segment formation itself. Thus, the *Fushi tarazu* gene, both on genetic and molecular grounds, appears to be correlated with the determination of segment formation, rather than the mechanisms of segment formation itself (Hafen et al. 1984). These ideas albeit consistent and logical are based on indirect evidence and analogy, and therefore must be viewed critically. However, they are valuable, since they define a model that can be subjected to experimental verification or rejection.

CONCLUSION AND FUTURE PROSPECTS

The demonstration of homeo-box affiliation with homeotic complexes and genes in *Drosophila* argues for an involvement of the homeo-box sequence in the genetic control of pattern formation. The conservation of homeo-box sequences in a broad range of metameric organisms suggests the conservation of such a function in other organisms as well. This is a tentative hypothesis still to be proven. If the hypothesis is correct, then there are major implications for our understanding of developmental processes in higher organisms.

The question facing us now is, what is the most efficient means to determine the function, if any, of the homeo-box sequences in higher forms? The existence of homeo-box sequences in *Mus musculus* provides a unique opportunity to answer this question, because of the recognized advantages of the laboratory mouse for molecular genetic and developmental studies. Of major importance will be the clear-cut demonstration of expression of the homeo-box sequences in mRNA and the characterization of a protein product, and then as a next step, a detailed developmental analysis of the temporal and spatial expression of the different homeo-box sequences. Beyond this, we must attempt to work out the structure of the homeo-box genes and their modes of control. Finally, if the homeo-box products are regulating the expression of other genes, this must be firmly established and the relevant genes identified. Much remains to be done, but if as we expect, the *Drosophila* homeo-box sequences will provide insight into vertebrate mechanisms of development, the effort will be well justified.

REFERENCES

Barker, P. E., F. H. Ruddle, H. D. Royer, O. Acuto, and E. L. Reinherz. 1984. Chromosomal location of human T cell receptor gene $T_1\beta$. *Science* **226**: 348.

Bender, W., M. Akam, F. Karch, P. A. Beachy, M. Peifer, P. Spierer, E. B. Lewis, and D. S. Hogness. 1983. Molecular genetics of the Bithorax complex in *Drosophila melanogaster. Science* **221**: 23.

Carrasco, A. E., W. McGinnis, W. J. Gehring, and E. M. DeRobertis. 1984. Cloning of an *X. laevis* gene expressed during early embryogenesis that codes for a protein domain homologous to *Drosophila* homeotic genes. *Cell* **37**: 409.

Green, M. C. (ed.). 1981. *Genetic variants and strains of the laboratory mouse.* Gustav Fischer Verlag, Stuggart, New York.

Hafen, E., A. Kuroiwa, and W. J. Gehring. 1984. Spatial distribution of transcripts from the segmentation gene *fushi tarazu* during *Drosophila* embryonic development. *Cell* **37**: 833.

Kaufman, T. C., R. Lewis, and B. Wakimoto. 1980. Cytogenetic analysis of chromosome 3 in *Drosophila melanogaster*: The homeotic gene complex in polytene chromosome interval 84A-B. *Genetics* **94**: 115.

Laughon, A. and M. P. Scott. 1984. Sequence of a *Drosophila* segmentation gene: Protein structure homology with DNA binding proteins. *Nature* **310**: 25.

Lee, N. E., P. D'Eustachio, D. Pravtcheva, F. H. Ruddle, S. M. Hedrick, and M. M. Davis. 1984. The beta chain of the murine T cell receptor is encoded on chromosome 6. *J. Exp. Med.* **160**: 905.

Levine, M., G. M. Rubin, and R. Tjian. 1984. Human DNA sequences homologous to a protein coding region conserved between homeotic genes of *Drosophila. Cell* **38**: 667.

Lewis, E. B. 1978. A gene complex controlling segmentation *Drosophila. Dev. Biol.* **102**: 565.

McGinnis, W., M. Levine, E. Hafen, A. Kuroiwa, and W. Gehring. 1984a. A conserved DNA sequence found in homeotic genes of the *Drosophila* Antennapedia and bithorax complexes. *Nature* **308**: 428.

McGinnis, W., R. Garger, J. Wirz, A. Kuroiwa, and W. Gehring. 1984b. A homologous protein-coding sequence in *Drosophila* homeotic genes and its conservation in other metazoans. *Cell* **37**: 403.

McGinnis, W., C. P. Hart, W. J. Gehring, and F. H. Ruddle. 1984c. Molecular cloning and chromosome mapping of a mouse DNA sequence homologous to homeotic genes of *Drosophila. Cell* **38**: 675.

Muller, M. M., A. E. Carrasco, and E. M. DeRobertis. 1984. A homeo-box containing gene expressed during oogenesis in *Xenopus. Cell* **39**: 157.

Ruddle, F. H. 1984. Reverse genetics and beyond. *Am. J. Hum. Genet.* **36**: 944.

Shepherd, J. C. W., W. McGinnis, A. E. Carrasco, E. M. DeRobertis, and W. J. Gehring. 1984. Fly and frog homeo domains show homologies with yeast mating type regulatory proteins. *Nature* **310**: 70.

Wakimoto, B. T. and T. C. Kaufman. 1984. Defects in embryogenesis in mutants associated with the *Antennapedia* gene complex of *Drosophila melanogaster. Dev. Biol.* **102**: 147.

Regulated Expression of a Foreign β-Globin Gene in Transgenic Mice

KIRAN CHADA, JEANNE MAGRAM, KATHRYN RAPHAEL,* GLENN RADICE, ELIZABETH LACY,† AND FRANK COSTANTINI
Department of Human Genetics and Development
College of Physicians and Surgeons
Columbia University
New York, New York 10032

OVERVIEW

To examine the *cis*-acting DNA sequences involved in the tissue-specific and temporal regulation of adult β-globin gene expression during mouse development, a hybrid mouse/human β-globin gene was introduced into the mouse germline. The foreign gene was transcribed specifically in erythroid tissues of adult animals in four different transgenic lines. In addition, the gene was silent in embryonic (day 10.5 or 11.5) erythroid cells but expressed in fetal liver erythroid cells, and thus was subject to temporal regulation similar to that of the endogenous adult β-globin genes. The steady-state level of mRNA from the foreign gene, however, was in no case greater than a few percent of the level of endogenous β-globin mRNA. These studies indicate that regulatory sequences closely linked to the adult β-globin gene are sufficient to establish the correct pattern of tissue-specific expression, and the correct pattern of developmental switching. More distant regulatory elements, absent from the microinjected DNA fragment, may be required for quantitatively normal levels of expression.

INTRODUCTION

The globin genes are subject to both tissue-specific and temporal regulation and therefore represent an interesting system for the study of developmental gene regulation in mammals. During embryonic, fetal, and postnatal development, different genes within the α- and β-globin gene clusters are sequentially expressed, a phenomenon known as hemoglobin switching. In the mouse, for example, the erythroid cells that are formed in the embryonic yolk sac between days 7 and 11 of gestation synthesize the embryonic β-like globin chains y and z. Only later in development, in the erythroid cells, which differentiate in fetal liver and in adult bone marrow, are the adult β-globin genes expressed (Russell 1979).

---

*Present address: CSIRO, Division of Animal Production, Ian Clunies Ross Animal Research Laboratory, Great Western Highway, Prospect, N.S.W., Australia
†Present address: Memorial Sloan-Kettering Cancer Center, New York, New York 10021

Although the structure and chromosomal organization of the globin genes in man and other vertebrates have been extensively characterized, virtually nothing is known about the DNA sequences within the globin gene clusters that are involved in the developmental regulation of transcription. The identification of regulatory sequences requires a functional assay; and gene transfer into the mouse germline via DNA microinjection into the zygote (Gordon et al. 1980; Wagner et al. 1981; Brinster et al. 1981; Costantini and Lacy 1981) provides a suitable assay system for regulatory sequences that operate during embryogenesis. Although several cloned α- and β-globin genes have been previously introduced into the mouse germline, these have not been expressed in an appropriate tissue-specific fashion (Stewart et al. 1982; Humphries et al. 1983; Lacy et al. 1983). We have recently observed, however, that a hybrid mouse/human β-globin gene can indeed be expressed in an erythroid-specific pattern in transgenic mice (K. Chada et al. 1985). In addition, the foreign globin gene is expressed in fetal liver, but not yolk sac-derived erythroid cells, and thus undergoes developmental switching appropriate for an adult β-globin gene in the mouse (Magram et al. 1985). Our observations, which are summarized in this paper, provide the first evidence that a cloned β-globin gene with a limited amount of flanking DNA contains sufficient *cis*-acting regulatory sequences to specify a correct tissue-specific and temporal pattern of expression during mouse development.

RESULTS

Introduction of a Hybrid Mouse/Human β-Globin Gene into the Mouse Germline

These experiments utilized a hybrid mouse/human β-globin gene, containing the 5′ portion of the mouse β-major globin gene and the 3′ portion of the human β-globin gene (plasmid pMHβ) (Chao et al. 1983). The first two-thirds of the mRNA coding sequence, the first intron, and 1.2 kb of 5′ flanking DNA were derived from the mouse β-major adult globin gene. The use of mouse DNA sequences ensured that any regulatory sequences within this region would be compatible with any species-specific regulatory factors in mouse erythroid cells. The 3′ end of the hybrid globin gene and the 3′ flanking region were derived from the human β-globin gene allowing the hybrid gene and its mRNA to be distinguished from the endogenous β-globin genes and their mRNAs. Previous studies (Chao et al. 1983) had shown that this hybrid gene could be expressed and regulated after transfection into mouse erythroleukemia cells, indicating that the cloned gene contains the sequence information for transcriptional activation during the late stages of erythropoiesis. Our experiments were intended to determine whether the same sequences are sufficient for tissue-specific and temporal regulation in the developing animal.

Mouse zygotes were microinjected with either the intact circular plasmid pMHβ, or with a purified linear DNA fragment containing the entire gene and 5′ flanking mouse DNA segment, 1.9 kb of 3′ flanking human DNA, and only 350 bp of plasmid DNA. Ten transgenic mice were produced, three carrying the entire plasmid (mice 11, 15, and 16), and seven carrying the purified DNA fragment (mice 43, 46, 47, 49, 53, and 77). When they were mated to normal mice, each of these transgenic mice transmitted the intact foreign gene to a fraction of its progeny, and these were bred to generate stable transgenic lines. Southern blot analysis demonstrated that animals in several of the transgenic lines contained multiple copies of the hybrid globin gene, apparently in tandem arrays (Costantini and Lacy 1981). Five of the lines which derive from zygotes injected with the purified DNA fragment carry only a single copy of the foreign gene per diploid genome (Fig. 1 and Table 1).

Specific Expression of the Hybrid β-Globin Gene in Erythroid Tissues

To ask whether the hybrid mouse/human β-globin gene was expressed in a tissue-specific fashion, RNA was isolated from various erythroid and nonerythroid tissues of adult mice in each line. Quantitative S1 nuclease analyses (Berk and Sharp 1977; Weaver and Weissman 1979) were performed, using a probe specific for human β-globin DNA sequences at the 3′ end of the gene (Chao et al. 1983), to measure the steady-state level of hybrid β-globin mRNA in each tissue.

Adult animals in four of the transgenic lines, all derived from embryos microinjected with the purified eukaryotic DNA fragment, were found to express the foreign globin gene primarily or exclusively in erythroid tissues. In these mice, the hybrid β-globin mRNA was detected at the highest levels in RNA isolated from whole blood and bone marrow (Fig. 2). Similarly high levels of the mRNA were detected in the spleen only after injection with phenylhydrazine (data not shown), a treatment which induces erythropoiesis in the spleen (Conkie et al. 1975). The hybrid β-globin mRNA synthesized was correctly initiated, spliced, and terminated, as shown by primer extension studies (not shown), as well as the 3′ S1 nuclease analyses (Fig. 2). A relatively minor amount of hybrid β-globin mRNA was also detected in several nonerythroid tissues in some of these animals. This was shown in most cases to result from blood contamination of the tissues, although in a few instances the mRNA appeared to represent actual expression of the foreign gene at a low level in a nonerythroid tissue.

The steady-state level of hybrid β-globin mRNA in each mouse RNA sample was measured relative to standard amounts of human β-globin mRNA (Fig. 2). In two of the lines showing tissue-specific expression (lines 46 and 49), the level of hybrid β-globin mRNA in reticulocytes was calculated to be approximately 700 molecules per cell. This corresponds to 1–2% of the level of endogenous β-major globin mRNA. Because the mice analyzed contained only one copy of the foreign β-globin gene per diploid genome, the level of expression per gene copy was about 2–4% of the normal level. Initial experiments to test whether the hybrid β-globin mRNA is

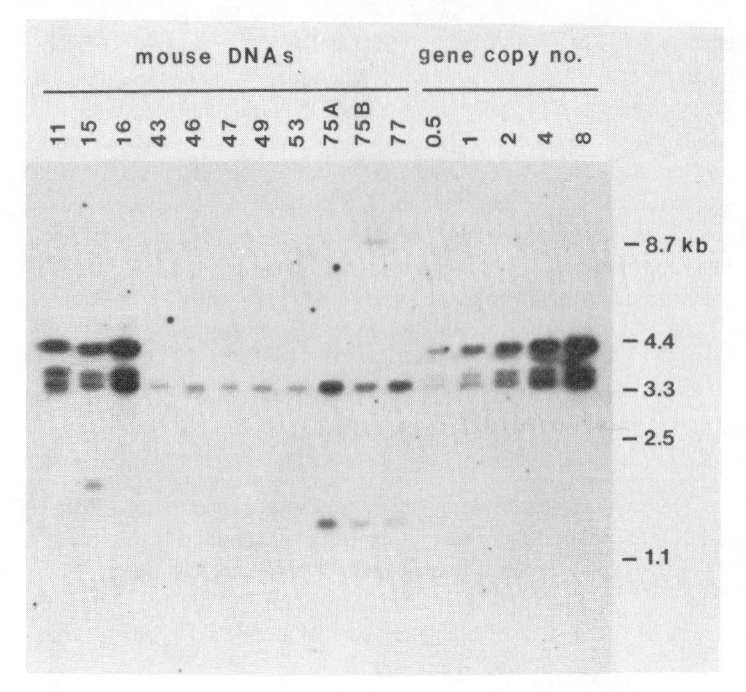

Figure 1
Detection of the hybrid mouse/human β-globin gene in transgenic mouse DNAs. DNA samples isolated from the tails of transgenic mice and plasmid pMHβ DNA were digested with the restriction endonuclease *Pst*I, fractionated by electrophoresis on a 0.7% agarose gel, and analyzed by Southern blot hybridization. This analysis shows that every mouse line carries at least one intact copy of the foreign gene and flanking DNA. The probe was the *Bam*HI–*Bgl*II fragment from the 3′ portion of the human β-globin gene cloned in a pBR322-derived vector. The first eleven lanes (marked *mouse DNAs*) contain tail DNAs from second or third generation mice in the indicated transgenic lines. All these lanes contain 5 μg of mouse DNA, except the first lane (mouse line 11), which contains 2.5 μg, and the third lane (mouse line 16), which contains 0.25 μg. The last five lanes contain plasmid pMHβ DNA in quantities corresponding to the indicated number of copies per diploid genome, per 5 μg of mouse DNA. Mouse lines 11, 15, and 16 derive from embryos injected with the entire pMHβ plasmid, and these mice show the same three *Pst*I fragments (4.2 kb, 3.6 kb, and 3.4 kb) as the plasmid DNA. The other transgenic mouse lines derive from embryos injected with the linear 5.0-kb *Cla*I–*Bgl*II fragment that contains the entire 3.4-kb *Pst*I fragment.

translated in blood cells of the transgenic mice were inconclusive, but they suggested that the hybrid β-globin polypeptide chain could account for no more than a few percent of the total β-globin. In hemizygous females from the other two lines that showed tissue-specific expression, 77 and 75, the observed mRNA levels relative to endogenous β-globin mRNA were approximately 0.1% and 0.02%, respectively.

The other three transgenic lines carrying the purified eukaryotic DNA fragment, as well as two of the lines carrying the entire plasmid pMHβ, showed no expression

Table 1
Transgenic Mouse Lines Carrying a Hybrid Mouse/Human β-Globin Gene

DNA injected	Mouse line	Gene copy number per diploid genome	Major site of expression
Entire pMHβ plasmid	11	9	None
	15	3	All tissues
	16	180	None
Purified 5-kb DNA fragment	43	1	None
	46	1	Erythroid tissues
	47	1	None
	49	1	Erythroid tissues
	53	1	None
	75	6	Erythroid tissues
	77	3	Erythroid tissues

of the foreign gene in any tissue tested. In the remaining line (15), which carried the entire plasmid, a very low level of hybrid β-globin mRNA was detected in non-erythroid as well as erythroid tissues (Table 1).

Temporal Regulation of the Foreign β-Globin Gene

In the mouse, the adult β-globin genes are first expressed on the twelfth day of gestation in the erythroid cells of the fetal liver, whereas the embryonic erythroid cells produced in the yolk sac at earlier stages synthesize only the embryonic β-like globin chains y and z (Russell 1979). We asked whether the hybrid mouse/human adult β-globin gene would be subject to the same pattern of temporal regulation by comparing its expression in the fetal liver and in the yolk sac-derived erythroid cells. Total RNA was isolated from circulating erythrocytes of 10.5-day and 11.5-day embryos, which derive from the yolk sac, and from fetal livers of 12.5-day and later stage fetuses; and hybrid β-globin mRNA was measured using the 3′ S1 nuclease assay. In two transgenic lines examined (46 and 77), expression was observed only in the fetal livers and not in the yolk sac-derived blood cells (data not shown). Thus a cloned adult β-globin gene in transgenic mice shows a pattern of activation during development similar to that of the mouse adult β-globin genes.

DISCUSSION

These studies have demonstrated that a 5-kb DNA fragment containing an adult β-globin gene and immediate flanking sequences includes all the *cis*-acting regulatory information needed to activate the gene specifically in erythroid cells during the development of a transgenic mouse. In addition, the cloned gene appears to contain the information for correct temporal regulation, or "hemoglobin switching,"

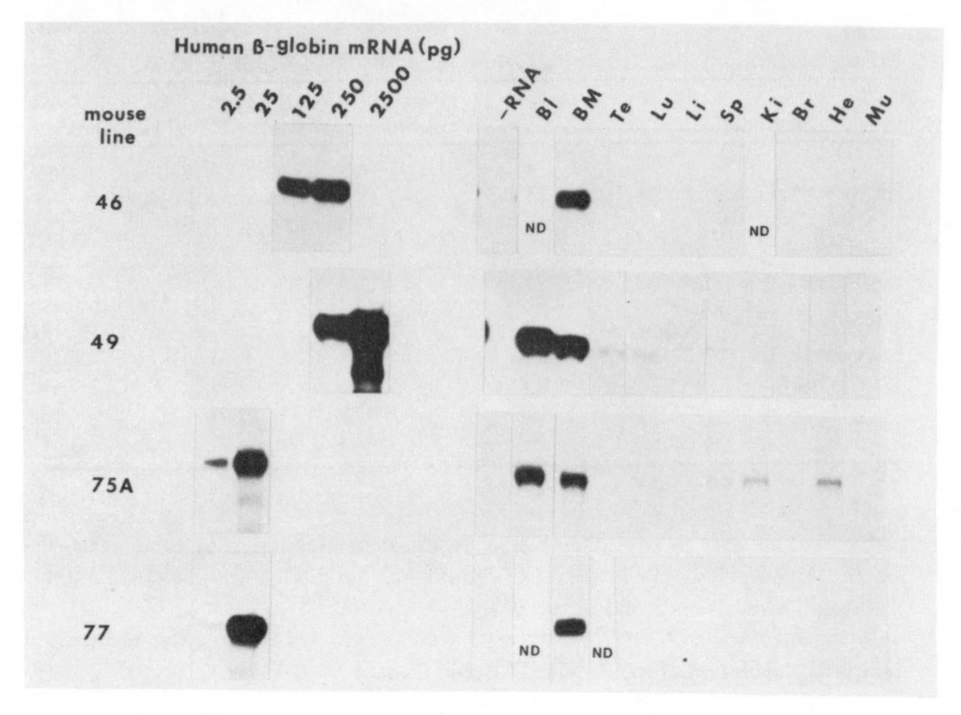

Figure 2
Specific expression of the hybrid mouse/human β-globin gene in erythroid tissues. The quantity of hybrid mouse/human β-globin mRNA was measured in total RNA samples from the indicated tissues of a mouse in lines 46, 49, 75, and 77. After hybridization of the RNAs with a ^{32}P-labeled DNA probe specific for the human portion of the hybrid mRNA, the hybrids were digested with S1 nuclease, fractionated by electrophoresis on an acrylamide gel and visualized by autoradiography. Only the regions of the autoradiograms showing the specific 212nt-labeled DNA fragment protected by hybrid mouse/human β-globin mRNA or human β-globin mRNA are shown. For each mouse analyzed, the hybridization signals were quantitated in relation to the signals at the left, obtained with known amounts of human β-globin mRNA. The intensity of hybridization signals from the four mice cannot be directly compared because the autoradiograms were exposed for different lengths of time. (*–RNA*) Hybridization with only yeast RNA carrier; (*Bl*) blood; (*BM*) bone marrow; (*Te*) testes; (*Lu*) lung; (*Li*) liver; (*Sp*) spleen; (*Ki*) kidney; (*Br*) brain; (*He*) heart; (*Mu*) skeletal muscle; (ND) not determined for the animal whose analysis is shown here. We used 10 μg of RNA from blood and 25 μg of RNA from all other tissues for each hybridization reaction. Reprinted, with permission, from Chada et al. (1985).

even when inserted at sites distant to the normal β-globin locus on mouse chromosome 7. The observation that the foreign gene was never expressed at a level equivalent to that of the endogenous β-globin genes, however, suggests that additional DNA sequences more distant from the adult β-globin gene, and absent from the microinjected DNA fragment, may be involved in some aspect of the gene's regulation. Alternatively, the low steady-state level of mRNA from the foreign gene may result from a low stability of the hybrid mRNA rather than from a low rate of transcription.

Not all transgenic mice carrying this gene were found to express it, suggesting that certain chromosomal integration sites may be completely incompatible with regulation of an inserted globin gene. The variation in levels of expression among the four transgenic mouse lines that express the gene in erythroid cells, and the trace levels of expression in a few nonerythroid tissues, may also reflect chromosomal position effects (Jaenisch et al. 1981; Lacy et al. 1983).

Given the present results, it is surprising that many other cloned α- and β-globin genes have failed to be expressed in erythroid tissues when introduced into the mouse germline (see Introduction). Although there are several differences between the cloned β-globin gene that we have utilized and the genes used in previous studies, we suspect that an important factor may have been the removal of plasmid vector sequences from the globin gene before microinjection (Table 1) (Chada et al. 1985).

Our results, as well as those of other studies described in this volume, suggest that the transgenic mouse system will be extremely useful for understanding the mechanisms by which genes are activated in specific cell types during development. Further studies in our laboratory will concentrate on the more precise localization of the *cis*-acting DNA sequences involved in the developmental regulation of the embryonic as well as the adult globin genes.

ACKNOWLEDGMENTS

K.C. was supported by a fellowship from S.E.R.C./NATO. F.C. is a recipient of an Irma T. Hirschl Career Scientist Award. This work was supported by grants to F.C. from the NIH and the National Foundation–March of Dimes.

REFERENCES

Berk, A.J. and P.A. Sharp. 1977. Sizing and mapping of early adenovirus mRNAs by gel electrophoresis of S1 endonuclease-digested hybrids. *Cell* **12**: 721.

Brinster, R.L., H.Y. Chen, M. Trumbauer, A.W. Senear, R. Warren, and R.D. Palmiter. 1981. Somatic expression of herpes thymidine kinase in mice following injection of a fusion gene into eggs. *Cell* **27**: 223.

Chada, K., J. Magram, K. Raphael, G. Radice, E. Lacy, and F. Costantini. 1985. Specific expression of a foreign β-globin gene in erythroid cells of transgenic mice. *Nature* **314**: 377.

Chao, M.V., P. Mellon, P. Charnay, T. Maniatis, and R. Axel. 1983. The regulated expression of β-globin genes introduced into mouse erythroleukemia cells. *Cell* **32**: 483.

Conkie, D., L. Kleiman, P.R. Harrison, and J. Paul. 1975. Increase in the accumulation of globin mRNA in immature erythroblasts in response to erythropoietin in vivo or in vitro. *Exp. Cell Res.* **93**: 315.

Costantini, F. and E. Lacy. 1981. Introduction of a rabbit β-globin gene into the mouse germ line. *Nature* **294**: 92.

Gordon, J., G.A. Scangos, D.J. Plotkin, J.A. Barbosa, and F.H. Ruddle. 1980. Genetic transformation of mouse embryos by microinjection of purified DNA. *Proc. Natl. Acad. Sci. U.S.A.* **77**: 7380.

Humphries, R.K., P. Berg, J. DiPietro, S. Bernstein, A. Baur, A. Nienhuis, and W.F. Anderson. 1983. Human and mouse globin gene sequences introduced into mice by microinjection of fertilized mouse eggs. In *Eucaryotic gene expression* (ed. A. Kumar, A.L. Goldstein, and G.V. Bahouney). Plenum Publishing, New York.

Jaenisch, R., D. Jahner, P. Nobis, I. Simon, J. Lohler, K. Harbers, and D. Grotkopp. 1981. Chromosomal position and activation of retroviral genomes inserted into the germ line of mice. *Cell* **24**: 519.

Lacy, E., S. Roberts, E.P. Evans, M.D. Burtenshaw, and F.D. Costantini. 1983. A foreign β-globin gene in transgenic mice: Integration at abnormal chromosomal positions and expression in inappropriate tissues. *Cell* **34**: 343.

Magram, J., K. Chada, and F. Costantini. 1985. Developmental regulation of a cloned adult β-globin gene in transgenic mice. *Nature* **315**: 338.

Russell, E.S. 1979. Hereditary anemias of the mouse: A review for geneticists. *Adv. Genet.* **20**: 357.

Stewart, T.A., E.F. Wagner, and B. Mintz. 1982. Human β-globin gene sequences injected into mouse eggs, retained in adults, and transmitted to progeny. *Science* **217**: 1046.

Wagner, E.F., T.A. Stewart, and B. Mintz. 1981. The human β-globin gene and a functional thymidine kinase gene in developing mice. *Proc. Natl. Acad. Sci. U.S.A.* **78**: 5016.

Weaver, R. and C. Weissman. 1979. Mapping of RNA by a modification of the Berk-Sharp procedure. The 5′ termini of 15S β-globin mRNA precursor and mature 10S β-globin mRNA have identical coordinates. *Nucl. Acid Res.* **7**: 1175.

Tissue-specific Expression of Immunoglobulin Genes in Transgenic Mice and Cultured Cells

RUDOLF GROSSCHEDL,* FRANK COSTANTINI,† AND DAVID BALTIMORE*
*Whitehead Institute for Biomedical Research
Cambridge, Massachusetts 02142
and
Department of Biology
Massachusetts Institute of Technology
Cambridge, Massachusetts 02139
†Department of Human Genetics and Development
College of Physicians and Surgeons
Columbia University
New York, New York 10032

OVERVIEW

Tissue-specificity of immunoglobulin (*Ig*) expression is maintained after transfer of a cloned rearranged μ heavy chain (H) gene into cultured cells or into the germline of mice. The transferred *Ig* gene is selectively transcribed in B and T lymphoid cells and is inactive in nonlymphoid cells. The expression of the *Ig* gene, when introduced into cultured cells, is dependent on a lymphocyte-specific transcriptional enhancer. The IgH enhancer can be replaced by a nonspecific viral enhancer without changing the tissue-specificity of gene expression. The IgH enhancer is therefore neither the only nor the dominant *cis*-acting regulatory element determining tissue-specific *Ig* gene expression.

INTRODUCTION

The differential activity of specific genes in different tissues is a central problem in developmental biology. What mechanism allows for the selective activation of a tissue-specific gene in the appropriate cell-type? What are the genetic elements which distinguish a tissue-specific gene from a nonspecific gene? Two approaches can be used to address these questions: One approach is the transfer of a cloned tissue-specific gene into the germline of an animal and the analysis of the expression of the "transgene" in the different tissues and cell types. The other approach is the introduction of the gene into cultured cell lines that were derived from various cell types. By comparing the phenotype of a tissue-specific wild-type gene with that of in vitro generated mutants in different cell types, it should be possible to identify *cis*-acting regulatory elements that control the tissue-specificity of gene expression.

Immunoglobulin (*Ig*) gene expression is tissue-specific in that it is restricted to lymphoid cells. These genes are encoded in multiple DNA segments that are joined together to generate functional immunoglobulin-coding genes. Is DNA rearrangement involved in the control of tissue-specificity of *Ig* gene expression? Are specific DNA sequences of a rearranged gene responsible for its selective expression in lymphoid cells?

RESULTS AND DISCUSSION

To test whether the rearranged *Ig* heavy chain gene is expressed in a tissue-specific fashion, we introduced a cloned rearranged mouse μ gene with the variable (V) region isolated from the hybridoma 17.2.25 (Loh et al. 1983) into the germline of mice (Grosschedl et al. 1984). The μ gene contained 2-kb 5′ flanking sequences and included the exons of both the secreted and membrane-bound form of the μ polypeptides (Fig. 1). We obtained five transgenic mice that harbored 20–140 copies of the introduced gene in their germline. The gene integrated in the germline will be called a transgene. Total RNA was isolated from various tissues of a second generation mouse. The RNA was analyzed by hybridization to a ^{32}P-labeled, single-stranded specific μ DNA probe followed by S1 nuclease digestion and electrophoretic separation. Faithful transcription was detected at a high level in spleen, lymph nodes, and thymus. No transcription of the transgene was detected in non-lymphatic tissues with the exception of heart which expressed the specific μ gene at

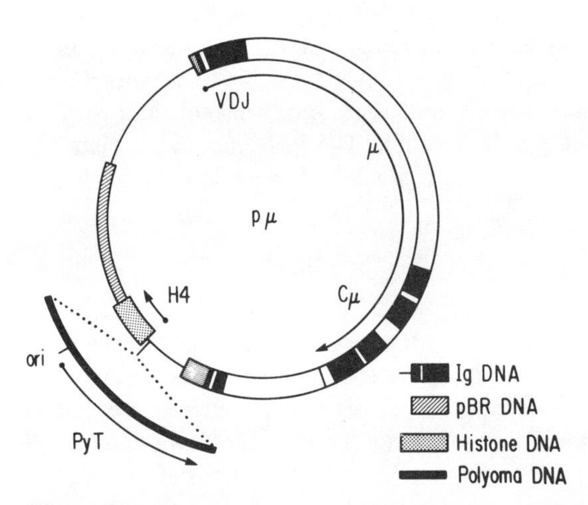

Figure 1
Structure of the plasmid pμ. The rearranged μ gene is shown as closed (exons), open (introns), and hatched (noncoding sequences) bars. The rearranged VDJ variable and the C constant regions are indicated. pBR322 vector sequences and H4 gene sequences are represented as hatched or dotted bars, respectively. The early region of polyoma (*Py*) virus containing the origin of replication (*ori*) and the large *T* gene is depicted as closed bar. The polyoma virus DNA segment is included in all DNA constructs which were used for short-term expression assays. The transcriptional polarities of the genes are indicated by arrows.

5% of the level detected in spleen (Grosschedl et al. 1984). From this result we concluded that the introduced rearranged μ gene is expressed in a tissue-specific fashion and that the introduced gene contains all *cis*-acting regulatory DNA sequences which are required for its lymphoid-specific expression. Furthermore, this result implies that the tissue-specific DNA rearrangement of *Ig* genes that occurs in B lymphoid cells does not solely account for the inactivity of the *Ig* gene in nonlymphoid cells. A similar conclusion was reached by the analysis of the expression of a rearranged κ light chain gene which had been introduced in the mouse germline (Brinster et al. 1983; Storb et al. 1984).

By fractionation of B and T lymphoid cells, we demonstrated that the introduced μ gene is almost equally active in both cell types (Grosschedl et al. 1984). This result indicated that the *cis*-acting regulatory elements of the heavy chain gene are functional in B and T cells, suggesting that the normal restriction of μ gene expression to B lymphoid cells may be determined exclusively by the B cell specificity of DNA rearrangement. The tissue-specificity (lymphoid versus nonlymphoid) of *Ig* gene expression is, however, not dependent on DNA rearrangement.

We determined the precise initiation site of transcription of the V_H 17.2.25 gene by Mung bean nuclease mapping (Green and Roeder 1980) of the 5′ ends of heavy chain transcripts from the hybridoma 17.2.25. Total RNA was hybridized to a single-stranded ^{32}P-labeled μ DNA probe. Treatment of the hybrids with Mung bean nuclease generated a single protected fragment (Fig. 2A, lane *5*). The size of this fragment, as determined by its relative migration to the "sequence-ladder" of the coding strand (lanes *1–4*), mapped the V_H 17.2.25 specific mRNA 5′-end 24 nucleotides downstream of a TATA sequence (Fig. 2B). We corrected the migration of the nuclease-generated DNA fragment by 0.5 nucleotides rather than by 1.5 nucleotides (Sollner-Webb and Reeder 1979) because the capped nucleotide of the mRNA is more likely an A nucleotide than an U nucleotide (Breathnach and Chambon 1981). Although digestion of the hybrids with S1 nuclease produced a series of protected fragments (Fig. 2, lanes *7* and *8*), S1 nuclease mapping was used for the RNA analyses of the transfection experiments.

To identify and to study the regulatory DNA sequences which allow for the tissue-specific expression of the *Ig* genes, we performed short-term assays of wild-type and mutant μ genes in cultured cells. To maximize transient expression of the transfected *Ig* gene, we included the early region of polyoma virus (Fig. 1). This viral DNA segment allowed for the replication of the DNA in mouse cells (Queen and Baltimore 1983) but did not interfere with the regulation of the *Ig* gene. To control for the phenotypic effects of sequence alteration in the μ gene we used transcription of a modified H4 histone gene residing in the DNA construct as an internal reference (Fig. 1). The modified H4 gene had a 50-bp deletion at its 5′ end to distinguish its transcripts from endogenous mouse H4 mRNA. The *Ig* gene construct was transfected into myeloma MPC11 cells and total RNA was isolated after 2 days. The RNA was analyzed for the presence of specific μ and H4 transcripts by

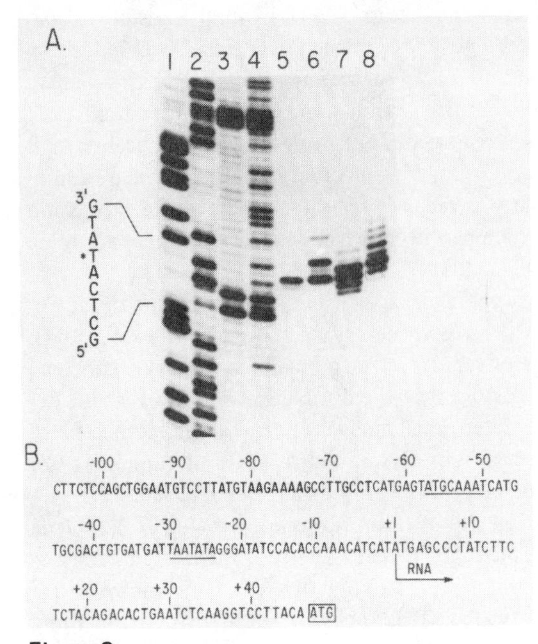

Figure 2
(*A*) High resolution nuclease mapping of the V_H 17.2.25 RNA 5′ terminus. The 17.2.25 hybridoma RNA was hybridized to the coding strand of the ^{32}P-labeled DNA probe and subjected to digestion with various amounts of either Mung bean nuclease or S1 nuclease. The protected fragments were separated next to a "sequence ladder" of the DNA probe for precise size determination. (Lanes *1-4*) G, A>C, C>T, C+T specific sequences reactions, respectively; (lane *5*) 1000 units Mung bean nuclease used for digestion; (lane *6*) 400 units Mung bean nuclease; (lanes *7-8*) 450 and 150 units S1 nuclease, respectively. (*B*) Nucleotide sequence of the 5′ flanking and 5′ noncoding region of the V_H 17.2.25 gene. Underlined sequences include two conserved sequence motifs: the octanucleotide box at –50 (Parslow et al. 1984) and the TATA box at position –30. The site of transcription initiation is indicated.

hybridization to ^{32}P-labeled, single-stranded μ and H4 DNA probes, respectively, followed by S1 nuclease digestion and electrophoretic separation (Fig. 3). Analysis of V_H 17.2.25 specific μ transcripts resulted in the generation of a series of protected fragments of about 57 nucleotides which comigrated with fragments protected by hybridoma 17.2.25 RNA (Fig. 4B, lanes *4* and *6*). Thus transcription of the transfected μ gene in myeloma cells is faithful. S1 nuclease analysis of transcripts from the H4 reference gene yielded a 99-nucleotides long, protected fragment (Fig. 3B and Fig. 4B, *bottom panel*). Transcription of this H4 reference gene was used in all experiments as control for the transfection efficiency and RNA recovery.

When lymphoid cells are propagated in culture, *Ig* heavy chain genes show a tendency to undergo deletions in the large intron between the variable (V) region

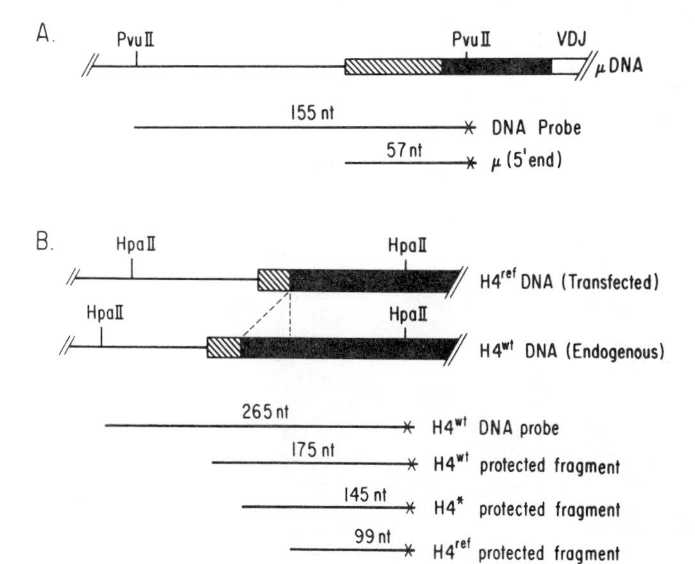

Figure 3
(*A*) Structure of the 5′ end-labeled μ DNA probe used for S1 nuclease mapping. The leader coding sequences are drawn as closed black boxes; intron sequences are shown as open and 5′ noncoding sequences as hatched bar. The structures and sizes of the single stranded DNA probe and the protected DNA fragment are shown. The position of the ^{32}P-label is indicated by an asterisk. (*B*) Structure of the 5′ end-labeled H4 DNA probe. The H4 coding sequences of the wild-type (wt) and modified reference (ref) genes are depicted as black boxes. The structure and sizes of the wt DNA probe which was used for S1 mapping and the protected fragments are shown. The modified reference gene transcripts generate a 99-bp protected fragment, H4 transcripts of the endogenous H4 wt or variant genes protect a 175-bp and 145-bp fragment, respectively.

and the constant (C) region. Some of these deletions lead to a reduction in the heavy chain expression (Alt et al. 1982) suggesting the existence of a regulatory element in the large intron. Indeed, functional tests of in vitro-generated deletions in the large intron of heavy chain genes demonstrated the existence of such *cis*-acting regulatory elements (Gillies et al. 1983; Neuberger 1983). We removed various DNA fragments from the rearranged μ gene and tested the mutants by a short-term transfection assay in myeloma MPC11 cells. Deletion of all $J_H/C\mu$ intron DNA sequences between the *Xba*$I_{1/3}$ sites (e.g., deletion mutant pμΔ1) (Fig. 4A) abolished detectable levels of μ gene transcription as assayed by S1 nuclease mapping of specific μ transcripts (Fig. 4B, lane *3*). The same "null-phenotype" was observed with deletion mutant pμΔ2 which lacked the 1-kb *Xba*$I_{1/2}$ DNA fragment (Fig. 4A; Fig. 4B, lane *2*). Analysis of the DNaseI sensitivity of the human heavy chain locus revealed a hypersensitive site downstream of the enhancer which in its sequence is conserved in the mouse μ heavy chain gene (Mills et al. 1983). We tested the putative func-

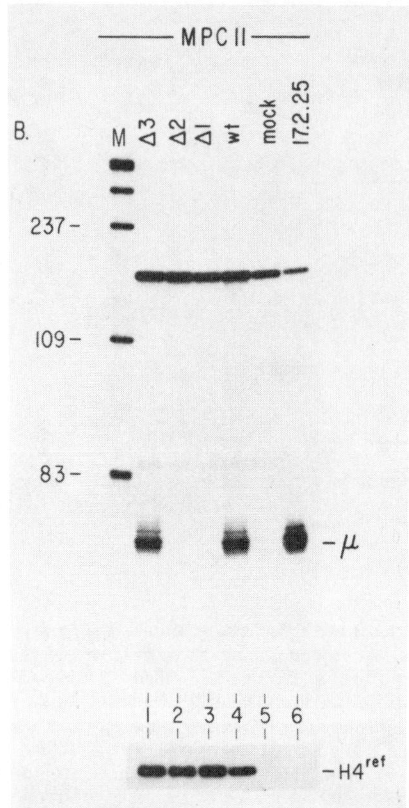

Figure 4B
Expression of wild-type and mutant μ genes in myeloma MPC11 cells. A total of 20 μg RNA from transfected cells were hybridized with the ^{32}P-labeled μ DNA probe and digested with S1 nuclease. The protected fragments were analyzed by electrophoresis through 8% polyacrylamide/urea gels. The position of the DNA fragment protected by specific μ transcripts is indicated. As control for transfection efficiency and RNA recovery, total RNA was hybridized with the ^{32}P-labeled H4 DNA probe and treated with S1 nuclease. The DNA fragment, which is protected by transcripts from the modified H4 gene, is shown in the bottom panel.

tional importance of this region by deleting the 3.3-kb *Xba*$I_{2/3}$ DNA fragment comprising the conserved sequence and the heavy chain gene switch-region between the enhancer of the Cμ exons (mutant pμΔ3) (Fig. 4A). This mutation directed μ gene transcription in MPC11 cells at wild-type levels (Fig. 4B, lane *1*), implying that the deleted DNA sequences are dispensable for heavy chain expression in myeloma cells. The RNA from the myeloma cells transfected with the various *Ig* gene constructs contained similar numbers of transcripts from the H4 reference gene (Fig. 4B, *bottom panel*) implying that the transfection efficiency was similar in each experiment.

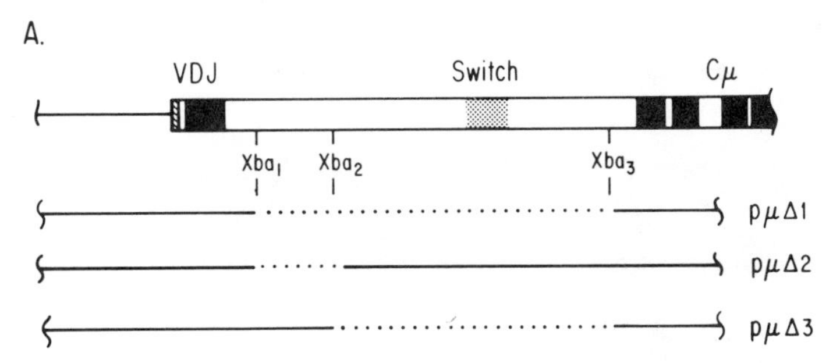

Figure 4A
Structure of μ gene deletion mutants. The μ gene is shown using the same symbols as in Figure 1. The three *Xba*I restriction sites used for the generation of deletions in the intron between the variable and the constant region are numbered. Deletions are indicated by dotted lines.

To determine whether the regulatory sequences located in the $V_H/C\mu$ intron behave like transcriptional enhancers, we inserted the 1-kb *Xba*I$_{1/2}$ DNA fragment into the *Hin*dIII site 1 kb 5′ to the promoter of the $\mu\Delta1$ gene in either orientation (Fig. 5A). Transfection of these DNA constructs, termed $\mu\Delta1(\mu E1)$ and $\mu\Delta1$ $(\mu E1^*)$ into myeloma MPC11 cells and S1 nuclease analysis of specific μ transcripts revealed the accumulation of μ transcripts at a level 50% of that observed with the wild-type template (Fig. 5B, lanes *2, 4,* and *5*). This result demonstrated that regulatory sequences residing within the 1-kb *Xba*I DNA fragment can stimulate μ gene transcription 5′ or 3′ of the promoter and thus have the properties of a tran-

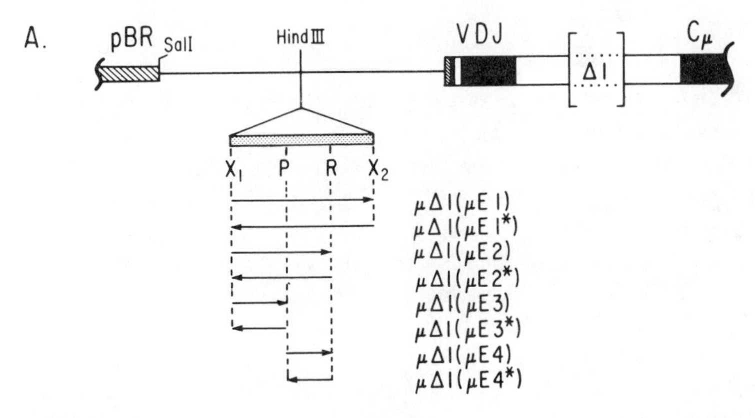

Figure 5A
Structure of enhancer insertion mutants. The $\mu\Delta1$ gene, which lacks its enhancer in the $J/C\mu$ intron, is shown. Various μ enhancer fragments were inserted in the *Hin*dIII site 5′ of the promoter. The enhancer fragments are drawn as arrows indicating their relative orientation to μ gene transcription. The restriction sites flanking the fragments are (*X*) *Xba*I, (*P*) *Pvu*II, and (*R*) *Eco*RI.

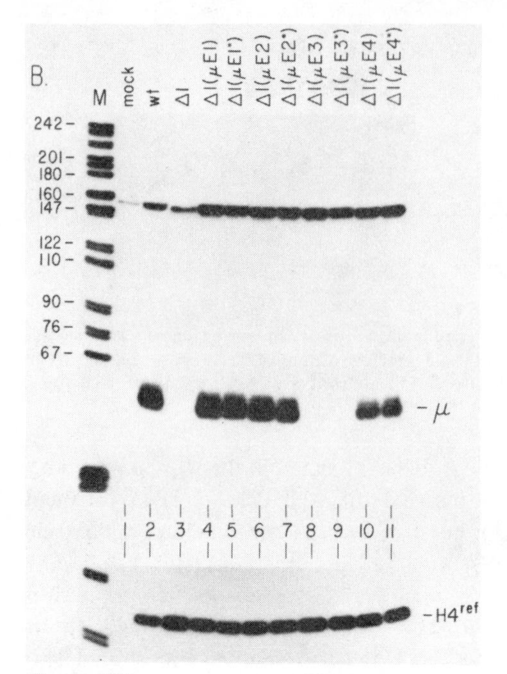

Figure 5B
Expression of the enhancer insertion mutants in myeloma MPC11 cells. Total RNA from cells transfected with various DNA constructs was analyzed by S1 nuclease mapping as described in Figure 4B. The position of the fragment protected by specific μ transcripts is shown.

scriptional enhancer (Khoury and Gruss 1983). This *Ig* enhancer is also required for transcription of stably transfected *Ig* genes (Gillies et al. 1983) and can activate heterologous genes (Banerji et al. 1983).

To localize the IgH enhancer in more detail, we dissected the 1-kb *Xba* DNA fragment in smaller DNA pieces. We tested the DNA fragments for enhancer activity by inserting them in either orientation 5′ to the promoter of the enhancerless μΔ1 gene (Fig. 5A). Transfection of the DNA constructs into myeloma MPC11 cells and analysis of the RNA by S1 nuclease mapping of specific μ transcripts revealed that the 700-bp *Xba*I/*Eco*RI fragment in the mutants μΔ1(μE2) and μΔ1(μE2*) has maximal enhancer activity (Fig. 5B, lanes *6* and *7*). The 400-bp *Xba*I/*Pvu*II DNA fragment was unable to enhance transcription in the mutant μ genes Δ1(μE3) and Δ1(μE3*) (Fig. 5B, lanes *8* and *9*) whereas the 300-bp *Pvu*II/*Eco*RI DNA fragment in mutants Δ1(μE4) and Δ1(μE4*) could activate μ gene transcription to a level 30% compared to the 700-bp *Xba*I/*Eco*RI fragment (lanes *10* and *11*). These data confirmed the location of the heavy chain enhancer to the *Pvu*II/*Eco*RI fragment as described by Banerji et al. (1983) but indicated the functional importance of additional specific DNA sequences in the *Xba*I/*Pvu*II fragment for maximal enhancer activity. Indeed, sequence comparison between the mouse and the human enhancer revealed striking sequence homology extending over 500 nucleotides within the *Xba*I/*Eco*RI fragment (Hayday et al. 1984; Mills et al. 1983).

The IgH enhancer can stimulate transcription of linked heterologous genes in lymphoid B cell lines but not in fibroblast cells (Banerji et al. 1983; Gillies et al. 1983; R. Grosschedl and D. Baltimore, unpubl.). This differential activity of the Ig enhancer formally explained the inactivity of a transfected rearranged heavy chain gene in fibroblast cells. If the Ig enhancer were the major *cis*-acting regulatory element determining tissue-specificity of gene expression, it would be possible to activate a rearranged heavy chain gene in nonlymphoid cells by replacing the IgH enhancer with a tissue-nonspecific enhancer. The enhancer of Moloney murine leukemia virus (Mo-enhancer) was demonstrated to be functional in fibroblast cells (Laimins et al. 1982). We inserted the Moloney virus enhancer 5′ to the promoter of the Ig enhancerless μ gene. Although this DNA construct directed μ gene transcription at wild-type levels in myeloma MPC11 cells, virtually no μ transcripts could be detected in transfected fibroblast cells (data not shown). A functional enhancer is obviously not sufficient for activation of the *Ig* heavy chain gene in fibroblast cells. The tissue-specificity of the *Ig* gene is not solely determined by the enhancer, but other *cis*-acting regulatory elements of the *Ig* gene control its lymphoid-specific transcription. More recent experiments have shown that the promoter region of the rearranged μ gene can confer tissue-specificity on a tissue-nonspecific heterologous gene. In addition μ intragenic sequences themselves contribute to the tissue-specific expression of *Ig* genes.

These results, which will be presented elsewhere, suggest that control of tissue-specific gene expression is complex and involves at least three regulatory elements: the enhancer, a region around the promoter, and one somewhere within the body of the gene itself. We are at present trying to locate the regulatory sequences and to determine at which level of gene expression they act.

ACKNOWLEDGMENT

This work was supported by grants from the National Institute of Health (HD17704), the March of Dimes (MDBDF5-410), and the American Cancer Society (IM-355P). R.G. is a postdoctoral fellow of the European Molecular Biology Organization. F.C. is a recipient of an Irma T. Hirschl Career Scientist Award.

REFERENCES

Alt, F.W., N. Rosenberg, R. Casanova, E. Thomas, and D. Baltimore. 1982. Immunoglobulin heavy chain class switching and inducible expression in an Abelson murine leukaemia virus transformed cell line. *Nature* **296**: 325.

Banerji, T., L. Olson, and W. Schaffner. 1983. A lymphocyte-specific cellular enhancer is located downstream of the joining region in immunoglobulin heavy chain genes. *Cell* **33**: 729.

Breathnach, R. and P. Chambon. 1981. Organization and expression of eukaryotic split genes coding for proteins. *Annu. Rev. Biochem.* **50**: 349.

Brinster, R.L., K.A. Ritchie, R.E. Hammer, R.L. O'Brien, B. Arp, and U. Storb. 1983. Expression of a microinjected immunoglobulin gene in the spleen of transgenic mice. *Nature* **306**: 332.

Gillies, S.D., S.L. Morrison, V.T. Oi, and S. Tonegawa. 1983. A tissue-specific transcription enhancer element is located in the major intron of a rearranged immunoglobulin heavy chain gene. *Cell* **33**: 717.

Green, M.R. and R.G. Roeder. 1980. Definition of a novel promoter for the major adenovirous-associated virus mRNA. *Cell* **22**: 231.

Grosschedl, R., D. Weaver, D. Baltimore, and F. Costantini. 1984. Introduction of a μ immunoglobulin gene into the mouse germ line: Specific expression in lymphoid cells and synthesis of functional antibody. *Cell* **38**: 647.

Hayday, A.C., S.D. Gillies, H. Saito, C. Wood, K. Wiman, W.S. Hayward, and S. Tonegawa. 1984. Activation of a translocated human c-*myc* gene by an enhancer in the immunoglobulin heavy-chain locus. *Nature* **307**: 334.

Khoury, G. and P. Gruss. 1983. Enhancer elements. *Cell* **33**: 313.

Laimins, L.A., G. Khoury, C. Gorman, B. Howard, and P. Gruss. 1982. Host-specific activation of transcription by tandem repeats from simian virus 40 and Moloney murine sarcoma virus. *Proc. Natl. Acad. Sci. U.S.A.* **79**: 6453.

Loh, D.Y., A.L. Bothwell, M.E. White-Scharf, T. Imanishi-Kari, and D. Baltimore. 1983. Molecular basis of a mouse strain-specific anti-hapten response. *Cell* **33**: 85.

Mills, F.C., L.M. Fisher, R. Kuroda, A.M. Ford, and H.J. Gould. 1983. DNaseI hyposensitive sites in the chromatin of human μ immunoglobulin heavy-chain genes. *Nature* **306**: 809.

Neuberger, M.S. 1983. Expression and regulation of immunoglobulin heavy chain gene transfected into lymphoid cells. *Eur. Mol. Biol. Organ. J.* **2**: 1373.

Parslow, T.G., D.L. Blair, W.J. Murphy, and D.K. Granner. 1984. Structure of the 5′ ends of immunoglobulin genes: A novel conserved sequence. *Proc. Natl. Acad. Sci. U.S.A.* **81**: 2650.

Queen, C. and D. Baltimore. 1983. Immunoglobulin gene transcription is activated by downstream sequence elements. *Cell* **33**: 741.

Sollner-Webb, B. and R.H. Reeder. 1979. The nucleotide sequence of the initiation and termination sites for ribosomal RNA transcription in *X. laevis*. *Cell* **18**: 485.

Storb, U., R.O. O'Brien, M.D. McMullen, K.A. Gollahon, and R.L. Brinster. 1984. High expression of cloned immunoglobulin κ gene in transgenic mice is restricted to β lymphocytes. *Nature* **310**: 238.

Expression of a Microinjected Immunoglobulin Kappa Gene in Transgenic Mice

URSULA STORB,* KINDRED A. RITCHIE,* ROBERT E. HAMMER,† REBECCA L. O'BRIEN,* JOANNA T. MANZ,* BENJAMIN ARP,* AND RALPH L. BRINSTER†

*Department of Microbiology and Immunology
University of Washington
Seattle, Washington 98195
†Laboratory of Reproductive Physiology
School of Veterinary Medicine
University of Pennsylvania
Philadelphia, Pennsylvania 19104

OVERVIEW

Transgenic mice, which have integrated a rearranged, functional immunoglobulin κ gene (pB1-14) in their somatic and germ cells, were produced. Six transgenic mice and their positive offspring have large amounts of the specific κ chains in their serum. Transcripts from pB1-14 were found at a high level in the spleen, but were undetectable in nonlymphoid tissues of testis, liver, kidney, heart, muscle, brain, and thyroid gland. In lymphoid cell subpopulations, the level of pB1-14 transcripts is correlated with the relative number of B cells; there is no correlation with the proportion of T lymphocytes. We conclude, therefore, that the microinjected κ gene contains target sequences for B lymphocyte-specific gene activation signals that override the influence of the integration site.

Hybridomas were produced from spleen cells of these κ transgenic mice to investigate expression of the transgenic κ gene, its effect on allelic exclusion, and its effect on the control of endogenous light chain gene rearrangement and expression. The results show that the transgenic gene is expressed normally and that the production of a complete immunoglobulin molecule turns off light chain gene rearrangement.

INTRODUCTION

We have made use of transgenic mice to dissect some parameters of immunoglobulin (*Ig*) gene expression. In this report, we concentrate on two aspects, the tissue-specific expression of *Ig* genes and the mechanism for allelic exclusion of *Ig* genes.

Transgenic mice have certain advantages for immunoglobulin gene studies. Compared with the analysis of polyclonal lymphocytes in normal mice, all the lymphocytes of a transgenic mouse carry the same rearranged gene. It may therefore be possible to analyze the response of that gene on the level of the whole animal. Because the transgenic lymphocytes undergo normal developmental and differentiation processes, they probably reflect normal immune responses, in contrast to myeloma tumor cells, which are more or less frozen in one particular differentiation state. Transgenic mice also have great advantages compared with cell transfection studies: Potential problems due to the transformed phenotype of cultured cells are avoided; the expression in different tissues that have the injected gene at the same location in the genome can be directly compared; and the injected gene goes through a normal ontogenetic development program.

Immunoglobulin (*Ig*) genes, in contrast to other known structural genes of vertebrates, require a rearrangement step before expression into antibodies is possible (Hozumi and Tonegawa 1976; Sakano et al. 1979; Seidman et al. 1979). The control of expression of these genes seems therefore even more complicated than the control of transcription in other cases. When a particular *Ig* gene locus becomes expressed, both alleles of the respective constant region (C) gene are activated (Storb et al. 1981; Storb and Arp 1983) similar to what seems to be the rule for other genes on autosomes. However, the *Ig* gene rearrangement on each of the allelic chromosomes is different (Lenhard-Schuller et al. 1978; Wilson et al. 1979; Coleclough et al. 1981). Either only one allele is rearranged and the other one remains in germline configuration, or both are rearranged, but in different ways. So far, no example is known where both alleles code for an Ig chain that is secreted by the cell. In most cases, where the other allele is rearranged, it shows a misalignment of the gene segments (Seidman and Leder 1980; Walfield et al. 1981), so that a faulty protein is encoded. In one case, κ chains were encoded by both alleles, however, one of the L chains had deleted two amino acids and was unable to associate with the H chain (Bernard et al. 1981; Kwan et al. 1981). These findings provided the molecular basis for the observation of "allelic exclusion" of *Ig* genes (Pernis et al. 1965). Obviously, in the interest of producing monospecific high affinity antibodies, a mechanism had developed to exclude the functional activation of more than one allele, which would cause scrambled antibody molecules. As the basis of this mechanism, two major possibilities have been considered: (1) There is a feedback from a functional *Ig* gene, or rather its protein product, to prevent further functional rearrangements in the same *Ig* locus (Alt et al. 1980; Bernard et al. 1981; Kwan et al. 1981); and (2) the chance for any functional rearrangement is extremely low, so that the chance for two such rearrangements in the same cell is minute (Walfield et al. 1980; Coleclough et al. 1981).

In order to determine if a functionally rearranged L chain gene influences other *Ig* gene rearrangements, we microinjected a functional κ gene (cloned from the myeloma MOPC-21) into fertilized mouse eggs and produced κ transgenic mice (Brinster et al. 1983; Storb et al. 1984). For the first time, in these mice, a functionally rearranged κ gene is present in all B cells before any endogenous *Ig* genes begin to rearrange. Hybridomas were made from spleen cells of these transgenic mice (Ritchie et al. 1984) to investigate the transgenic κ gene's expression on a single cell level, its effect on allelic exclusion, and its effect on the control of light chain gene rearrangements.

RESULTS

Characterization of the Microinjected DNA

The rearranged κ gene used for microinjection is shown in Figure 1. Six transgenic mice were produced (Brinster et al. 1983). All mice carry multiple copies (16–64) of the microinjected pB1-14 gene (Table 1). As in transgenic mice carrying other genes, most of the multiple copies are integrated at a single integration site in tandem head-to-tail fashion. The integrations are apparently stable because about 50% of the offspring of transgenic mice carry pB1-14 and the copy number per integration site appears to be the same as in the parents. As determined by Southern blots, the site of integration into host DNA is different in the six mice. One of the original transgenic mice (no. 54) has two different integration sites, name-

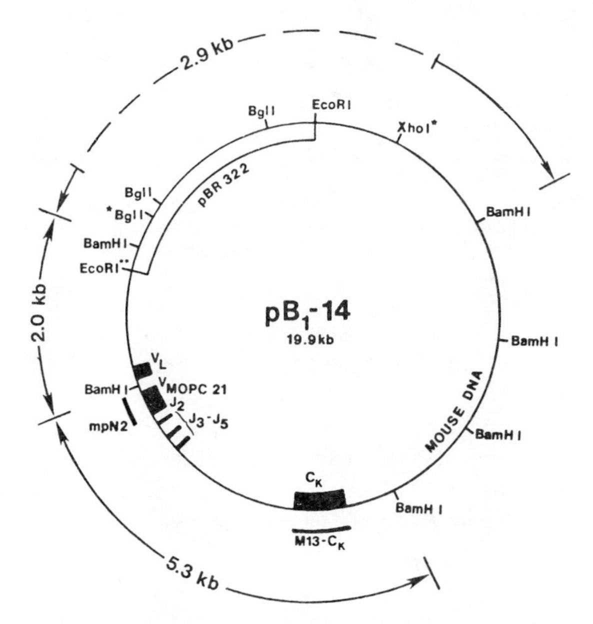

Figure 1
Map of plasmid DNA pB1-14. The V_{κ}M.21-C_{κ} gene insert of a phage clone (Walfield et al. 1980) was excised at the *Eco*RI sites and cloned into the *Eco*RI site of pBR322. The large *Bgl*I*-*Xho*I* fragment (14.8 kb) was separated from the smaller *Bgl*I-*Bgl*I and *Bgl*I-*Xho*I fragments and injected into mouse ova (Brinster et al. 1983). (V_L) Leader sequences; ($V_{MOPC\text{-}21}$) variable region; (J_2 to J_5) joining segments; (C_{κ}) constant region. Probes for V_{κ}M.21 (mpN2) and C_{κ} (M13-C_{κ}) are indicated. The broken line denotes the portion (5.9 kb) that was eliminated from the plasmid DNA before microinjection. Modified from Brinster et al. (1983).

Table 1
Relative Amounts of V_κM.21 RNA in Spleen of Different Transgenic and Normal Mice

Mice[a]	No. of copies of pB1-14 gene[b]	pg κRNA[c]/ μg total RNA	Percent[d] V_κM.21
SB (5F, 4 months)	–	19	1.7
Old C57BL/6 (2F, 10 months)	–	18	0.67
47 (1F, 2M, 5 1/2 months)	24	254[e]	32
48 (2F, 1-1/2 months)	48	18	21
49 (2F, 3 months)	16	27	40
52 (3F, 2M, 3 months)	32	17	25
54 (3F, 4 months)	64, 24[f]	27	28
55 (1M, 7 months)	32	96	25.8

[a]In parenthesis number, sex, and age of the mice; in all cases, except 55, the spleens from several mice were pooled before RNA extraction. Numbers 47–54 are F_1 offspring of transgenic mice. Number 55 is a first generation transgenic mouse. (SB) Normal SJL/J and C57BL/6 mice.

[b]The copy number of the microinjected gene in transgenic mice was estimated by dot hybridization of kidney DNA (47–54) or tail DNA (55).

[c]Approximate quantities of total κRNA and V_κM.21 containing RNA were determined from dot hybridizations in probe excess, relative to a cloned DNA standard. After exposure of the dot hybridizations to x-ray film (not shown), the dots were cut out and counted in a liquid scintillation counter.

[d]Percent of total κRNA which contains the V_κM.21 sequence.

[e]We have no explanation for the very high level of RNA in these number 47 mice. Three other number 47 mice were analyzed individually at 3 weeks of age and found to contain levels of κRNA in the range which is normal for other mice.

[f]The original transgenic mouse 54 has two insertion sites which segregate in offspring; both types of offspring transcribe the microinjected gene in the spleen, but not the liver. Of the three mice shown here, two had the low and one the high copy number. Modified from Storb et al. (1984).

ly in chromosomes 2 and 11 (Fig. 2). Chromosome two contains 64; chromosome eleven 24, copies of pB1-14. Mouse 54 is probably a mosaic, since individual chromosome spreads show hybridization to either chromosome, but not to both. Also the percentage of positive offspring of mouse 54 is only about 50%, and not 75% as expected if each germ cell contained both integrations. pB1-14 DNA from the two sites segregates independently in the offspring of mouse 54 (see Fig. 2 and Table 1). We therefore have seven different transgenic integrations.

Serum κ Chains Encoded by the Microinjected κ Gene

The MOPC-21 myeloma, from which pB1-14 was isolated, produces κ chains of relatively high molecular weight (Storb et al. 1980). Because of the unique V_κM.21 region, the κ chains have a distinct isoelectric point (Brinster et al. 1983). The protein is apparently glycosylated and produces two spots of characteristic size, shape, and location in 2-D protein gels (Fig. 3). All transgenic mice carrying the pB1-14 gene show the two characteristic spots in 2-D gels of their serum. Sera from normal littermates do not show these spots.

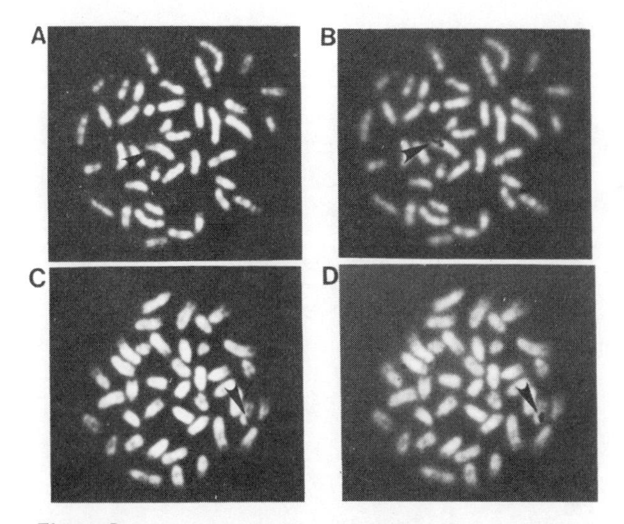

Figure 2
Localization of pB1-14 sequences in metaphase chromosomes of transgenic mice after in situ hybridization with ^{3}H-labeled pBR322 probe. (*A*) and (*C*) show Q-banded metaphase chromosomes from bone marrow of transgenic mouse 54 and one of its offspring, respectively. The chromosomes in (*A*) and (*C*) were visualized under fluorescent light. (*B*) and (*D*) show the same spreads visualized under a combination of fluorescent light and visible light. A cluster of silver grains can be seen on the distal end of chromosome 2 in (*B*) and on the proximal end of chromosome 11 in (*D*) (see text for details). In situ hybridization was performed using metaphase spreads obtained directly from bone marrow (Disteche et al. 1979) of transgenic mice. The hybridization procedure is essentially as described (Harper and Saunders 1981) except that the hybridization temperature was 42°C and washing was done at 48°C. pBR322 was nick translated with three tritiated nucleotides to a specific activity of 3.0×10^7 cpm/μg. Following autoradiography, chromosomes were stained with quinacrine mustard (Casperson et al. 1968) and subsequently identified using the standard nomenclature for the mouse (Nesbitt and Francke 1973).

Expression of the Microinjected κ Gene into RNA

Transcription of the microinjected pB1-14 gene can be monitored by using a DNA probe corresponding to the V region of pB1-14 (Fig. 1; mpN2). The particular V region (V_κM.21) is apparently rearranged and expressed only in a small percentage of B lymphocytes of normal mice. Approximately 0.7-4% of the κ chains in normal mouse serum carry this V region (Rokhlin et al. 1983). Similarly, only 0.67-3.3% of the κRNAs of normal spleens come from the pB1-14 related gene (Storb et al. 1984). It was surprising, then, to find that in all the transgenic mice which carry the microinjected pB1-14 gene, 20–40% of the RNAs of the spleen carry the V_κM.21 sequence (Table 1). The RNA is of normal size (Brinster et al. 1983).

The microinjected pB1-14 gene is expressed in normal tissue-specific fashion in all our transgenic mouse lines, i.e., in seven out of seven different integrations and in all positive offspring (Storb et al. 1984). Only the RNA of the spleen contains a

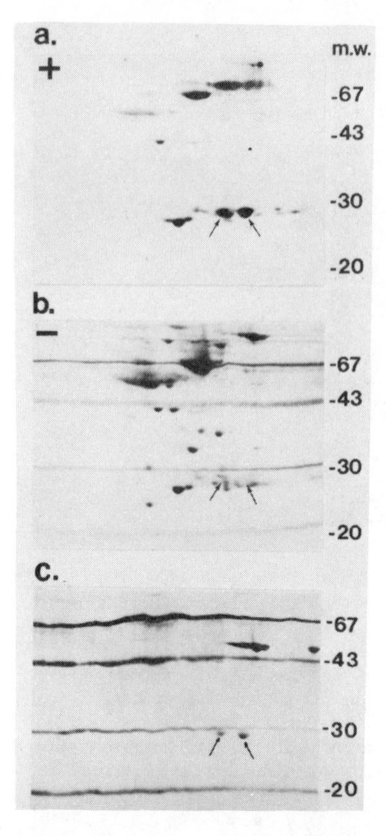

Figure 3
Two-dimensional gels of serum proteins (*a*) and (*b*), 25 μl of serum of κ-transgenic mouse (+) and normal littermate (–). The sera were precipitated with 40% ammonium sulphage, the precipitate redissolved in glass distilled water and run on 2-D gels (O'Farrell 1975), which were stained with Coomassie blue. (*c*) Purified MOPC-21 immunoglobulin (Litton Bionetics). Molecular weights (*m.w.*) of marker proteins run in the second dimension of the same gel (*b, c*) or of a control gel (*a*) are indicated.

large amount of pB1-14 encoded RNA (Table 1). The RNAs of liver, kidney, heart, muscle, brain, thyroid, and testis do not contain this RNA at a level detectable by dot hybridization. Of course, all organs contain a small amount of immunoglobulin RNA due to lymphocytes from circulating blood. For example, liver of both normal and transgenic mice contains 0.6–0.9% of the κRNA levels found in the spleen (Brinster et al. 1983).

Small amounts of pB1-14 transcripts were found in the thymus and bone marrow of transgenic mice. Again, the total level of κRNA in these organs in transgenic mice is not higher than in normal mice. Both thymus and bone marrow contain some B lymphocytes. Spleen cells enriched for B cells contain large amounts, spleen cells enriched for T cells only small amounts of the transgenic transcripts (Storb et al. 1984). Certain Abelson virus transformed pre-B cell lines from transgenic mice carrying the pB1-14 κ gene produce μmRNA, but not the transgenic (or other) κmRNA (Storb et al. 1985). The microinjected κ gene appears, therefore, to be expressed in tissue and development-specific fashion–only in mature B lymphocytes.

Production of Hybridomas from κ Transgenic Mice

Hybridomas were made from the spleens of κ transgenic mice (Ritchie et al. 1984). Hybridomas were first tested for pBR322 DNA, which was included on the microinjected κDNA to serve as a marker for the presence of the injected gene (data not shown). For further analysis, 24 clones retaining the microinjected gene and two clones that had probably lost the chromosome bearing the microinjected gene were selected. Each of the 26 hybridoma clones was analyzed for the production of Ig mRNA, Ig protein, and rearrangement of *Ig* heavy and light chain genes.

The following is a summary of the analysis of the various types of hybridomas (Table 2) (it should be noted that the transgenic κ protein cannot be secreted unless coupled to a heavy chain): (1) In transgenic κ secretors, different heavy chains are made, the transgenic κ chain associates with heavy chain and is secreted in a complete immunoglobulin molecule; no endogenous light chains are detected and no endogenous light chain genes are rearranged. (2) In endogenous κ secretors that secrete free light chain, heavy chain mRNA is present, but is apparently nonfunctional, since no heavy chains are made; the transgenic κ chain is present in the cytoplasm but not secreted; endogenous κ genes are rearranged. (3) In endogenous κ secretors that secrete whole Ig, heavy chain proteins are made, the transgenic gene is present but apparently not functional, since no transgenic κ protein is found; endogenous κ genes are rearranged, expressed into protein, and assembled into a complete immunoglobulin molecule. (4) In four nonsecreting hybridomas, heavy chains have not been detected, the transgenic κ protein is present in the cytoplasm but not secreted, no endogenous κ chains are found in the cytoplasm, but endogenous κ genes have been found in either a germline or rearranged context.

It is important to note cell types that have not been detected:

1. Transgenic κ secretors with cytoplasmic endogenous κ chains; suggesting that in cells where the microinjected gene contributes to a functional immunoglobulin molecule, endogenous κ chains are not produced.
2. Hybridomas that secrete both transgenic and endogenous κ chains, i.e., allelic exclusion is observed.
3. Hybridomas that secrete λ light chains. It is known that κ transgenic mice have normal amounts of λmRNA in their spleen and λ chains in their serum (S. Ogden, K.A. Ritchie, and U. Storb, unpubl.), but a λ secreting hybridoma has

Table 2
Summary of Hybridomas

Type	Cytoplasmic κ chains		Secreted κ chains		pBR322 DNA present	Endogenous κ genes rearranged	H-chain mRNA		H-chain Protein		Number tested for H-protein	Numbers
	Trans-genic	Endog-enous	Trans-genic	Endog-enous			μ	γ	μ	γ		
Transgenic[a]												
κH secretors	+	–	+	–	+	–	4	7	4	6	10/11	11
Endogenous[b]												
κH secretors	–	+	–	+	+	+	2	0	2	0	2/2	2
Endogenous[c]												
free κ secretors	+	+	–	+	+	+	5	2	–	–	7/7	7
Nonsecretors[d]	+	–	–	–	+	2+ 2–	2	2	–	–	4/4	4
Controls[e]	–	–	–	–	–	+	1	1	–	–	2/2	2

Summary table of the different kinds and numbers of hybridomas that have been analyzed from κ transgenic mice. Cytoplasmic and secreted κ chains were analyzed as in Figure 1 and 2; RNA as described in the text. The presence of pBR322 DNA was detected by preparing DNA from hybridomas, dotting onto nitrocellulose and hybridizing with radioactively labeled pBR322 (data not shown).

[a] One hybridoma from mouse 48, three from mouse 54, and seven from mouse 55
[b] One from mouse 48, one from mouse 55
[c] Two from mouse 48, four from mouse 54, and one from mouse 55
[d] One from mouse 54, and three from mouse 55
[e] One from mouse 48, and one from mouse 54
(+) Present; (–) absent
Reprinted, with permission, from Ritchie et al. (1984).

not been found perhaps because not enough hybridomas have yet been characterized.

These data are most consistent with the hypothesis (Alt et al. 1980; Bernard et al. 1981; Kwan et al. 1981) that a complete, functional immunoglobulin (not just a functional light chain) plays a role in turning off rearrangement. A model to explain these observations is outlined in Figure 4.

DISCUSSION

The microinjected pB1-14 κ gene is expressed at a high rate in B lymphocytes, but not in any other tissue. In transgenic mice produced with other genes often no expression (or variable expression) was found (Brinster et al. 1981; Jaenisch et al. 1981; Wagner et al. 1981; Lacy et al. 1983; Palmiter et al. 1983). Nonexpression may be due either to the presence in the cloned microinjected genes of inhibitory sequences or to insertion into a normally inactive genomic region, coupled with the absence of a strong positive regulatory element in the microinjected gene. Although

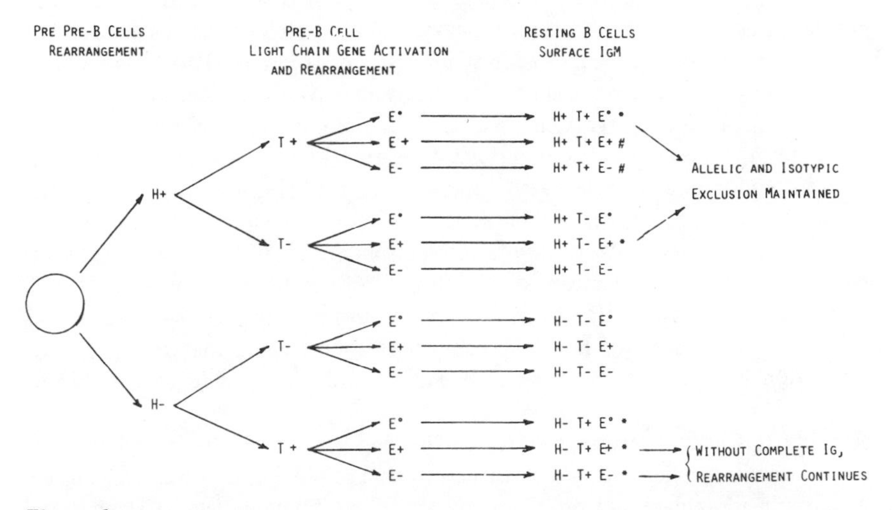

Figure 4
A scheme to explain the various types of transgenic hybridomas within the context of normal B cell development. Functionality is defined by the ability to be part of a complete Ig molecule. (*H*) Heavy chain; (*T*) transgenic chain. The T– symbol does not mean that the transgenic κ genes rearrange. Rather, T– represents the chance occurrence that a particular B cell happens to inactivate the transgenic κ gene by some event. (E) Endogenous light chain genes. Along the normal program of B cell development, a transgenic B cell has to undergo three major events: Rearrange a heavy chain gene; activate or, by chance, inactivate the transgenic gene; and finally, rearrange (or not rearrange) endogenous light chain genes. Options at each of these events will produce the various types of B cells on the right. Allelic and isotypic exclusion is the result of forming a complete antibody molecule. (*o*) Unrearranged; (+) functionally rearranged; (–) nonfunctionally rearranged; (*) cell types found; (#) cell types not found, i.e., rearrangement stops. Reprinted, with permission, from Richie et al. (1984).

these possibilities have not been formally distinguished, we favor the notion that the pB1-14 κ gene is expressed in such a highly tissue and development-specific manner because of positive regulatory elements present in the microinjected gene (Storb et al. 1984). It appears likely that the cloned κ gene is inactivated in the fertilized egg and remains so in all cells. In the B cell lineage, it seems to become transcriptionally active only after endogenous κ genes have been rearranged and transcribed (Ritchie et al. 1984). Activation of the microinjected κ gene at an insertion site that may be inactive in normal B lymphocytes (Storb et al. 1984) probably requires restructuring of chromatin before transcription factors can gain access. Perhaps this opening up of chromatin is different from the activation of naked DNA transfected into an already highly differentiated B lymphocyte (Rice and Baltimore 1982; Gillies et al. 1983; Ochi et al. 1983; Oi et al. 1983; Falkner and Zachau 1984). It is our hope that transgenic mice carrying a cloned immunoglobulin gene may provide a model to study gene activation in development.

With respect to developmental and immunological characteristics so far assessed, the pB1-14 gene behaves normally. Despite the presence of this functionally rearranged κ gene in every B lymphocyte, phenotypic "allelic exclusion" is maintained (Fig. 4). The isolation of $H^+T^+E^\circ$ cells but the lack of $H^+T^+E^+$ and $H^+T^+E^-$ cell types indicates that when a functional heavy chain is present, the transgenic κ gene prevents endogenous light chain gene rearrangement. Finding $H^+T^-E^+$ type cells shows that when the transgenic gene is by chance inactivated, endogenous genes can indeed rearrange. Finally, $H^-T^+E^+$ and $H^-T^+E^-$ cell types show that without a heavy chain, the presence of the functional transgenic κ chain alone is not sufficient to prevent endogenous *Ig* gene rearrangement. The results strongly support the hypothesis (Alt et al. 1980; Bernard et al. 1981; Kwan et al. 1981) that the presence of a complete Ig molecule turns off rearrangement. This establishes a basis for allelic exclusion. The mechanism of this turning off is currently being investigated. It appears, since different H chains are present in the transgenic κ-secreting cells (not shown), that any two H and L chains can combine and that any H-L chain combination is capable of exerting this feedback inhibition of *Ig* gene rearrangement.

Despite this drastic feedback effect on the single cell level, the transgenic mice are not immunodeficient. This may in part be due to the fact that heavy chains are generally the dominant antigen-binding part of antibody molecules (Haber and Richards 1966) so that antibodies containing endogenous H chains together with the transgenic κ chain would be reasonably functional. More important, though, is probably the population of normal antibody molecules that are produced by endogenous κ secretor cells which by chance have inactivated the transgenic κ gene and secrete antibody molecules that are entirely formed by endogenous H and L chains. The reason for the inactivation of the microinjected gene in some cells is unknown. In one case (K.A. Ritchie and U. Storb, unpubl.), the multiple transgenic genes had become grossly rearranged as judged by Southern blots. Perhaps the whole cluster of microinjected genes can be excised and translocated into a genomic region which is incompatible with transcription. Alternatively, only one gene of the cluster may be transcribed to begin with, thus increasing the probability of inactivation by random mutations.

ACKNOWLEDGMENTS

This research was supported by NIH Grants CA/AI 25754, CA 35979, DE 02600, and HD 17321, and NSF Grant PCM 81-07172. K.A.R. was supported by NIH-NIGMS NRSA No. GM-07266; R.L.O'B. by an NIH predoctoral training grant; and J.T.M. by a stipend from Genetic Systems Corporation.

REFERENCES

Alt, F. W., V. Enea, A. L. M. Bothwell, and D. Baltimore. 1980. Activity of multiple light chain genes in murine myeloma cells producing a single, functional light chain. *Cell* **21**: 1.

Bernard, O., N. M. Gough, and J. M. Adams. 1981. Plasmacytomas with more than one immunoglobulin κmRNA: Implications for allelic exclusion. *Proc. Natl. Acad. Sci. U.S.A.* **78**: 5812.

Brinster, R. L., H. Y. Chen, M. Trumbauer, A. W. Senear, R. Warren, and R. D. Palmiter. 1981. Somatic expression of herpes thymidine kinase in mice following injection of a fusion gene into eggs. *Cell* **27**: 223.

Brinster, R. L., K. A. Ritchie, R. E. Hammer, R. L. O'Brien, B. Arp, and U. Storb. 1983. Expression of a microinjected immunoglobulin gene in the spleen of transgenic mice. *Nature* **306**: 332.

Casperson, T., S. Farber, G. E. Foley, V. Kudynowski, E. J. Modest, E. Simonsson, V. Wagh, and L. Zeck. 1968. Chemical differentiation along metaphase chromosomes. *Exp. Cell Res.* **49**: 219.

Coleclough, C., R. Perry, K. Karjalainen, and M. Weigert. 1981. Aberrant rearrangements contributed significantly to the allelic exclusion of immunoglobulin gene expression. *Nature* **290**: 372.

Disteche, C. M., E. M. Eicher, and S. A. Latt. 1979. Late replication in an X-autosome translocation in the mouse: Correlation with genetic inactivation and evidence for selective effects during embryogenesis. *Proc. Natl. Acad. Sci. U.S.A.* **76**: 5234.

Falkner, F. G. and H. G. Zachau. 1984. Correct transcription of an immunoglobulin κ gene requires an upstream fragment containing conserved sequence elements. *Nature* **310**: 71.

Gillies, S. D., S. L. Morrison, V. T. Oi, and S. Tonegawa. 1983. A tissue-specific transcription enhancer element is located in the major intron of a rearranged immunoglobulin heavy chain gene. *Cell* **33**: 717.

Haber, E. and F. F. Richards. 1966. The specificity of antigenic recognition of heavy chain. *Proc. Roy. Soc. Lond. B* **166**: 176.

Harper, M. E. and G. F. Saunders. 1981. Localization of single copy DNA sequences on G-banded human chromosomes by *in situ* hybridization. *Chromosoma (Berl.)* **83**: 431.

Hozumi, N. and S. Tonegawa. 1976. Evidence for somatic rearrangement of immunoglobulin genes coding for variable and constant regions. *Proc. Natl. Acad. Sci. U.S.A.* **73**: 3628.

Jaenisch, R., D. Jahner, P. Nobis, I. Simon, J. Lohler, K. Harbers, and D. Grotkopp. 1981. Chromosomal position and activation of retroviral genomes inserted into the germ line of mice. *Cell* **24**: 519.

Kwan, S.-P., E. E. Max, J. G. Seidman, P. Leder, and M. D. Scharff. 1981. Two kappa immunoglobulin genes are expressed in the myeloma S107. *Cell* **26**: 57.

Lacy, E., S. Roberts, E. P. Evans, M. D. Burtenshow, and F. D. Costantini. 1983. A foreign β-globin gene in transgenic mice: Integration at abnormal chromosomal positions and expression in inappropriate tissues. *Cell* **34**: 343.

Lenhard-Schuller, R., B. Hohn, C. Brack, M. Hirama, and S. Tonegawa. 1978. DNA clones containing mouse immunoglobulin κ chain genes isolated by *in vitro* packaging into phage λ coats. *Proc. Natl. Acad. Sci. U.S.A.* **75**: 4709.

Nesbitt, M. N. and U. Francke. 1973. A system of nomenclature for band patterns of mouse chromosomes. *Chromosoma* **41**: 145.

Ochi, A., R. G. Hawley, M. J. Shulman, and N. Hozumi. 1983. Transfer of a cloned immunoglobulin light-chain gene to mutant hybridoma cells restores specific antibody production. *Nature* **302**: 340.

O'Farrell, P. H. 1975. High resolution two-dimensional electrophoresis of proteins. 1975. *J. Biol. Chem.* **250**: 4007.

Oi, V. T., S. L. Morrison, L. A. Herzenberg, and P. Berg. 1983. Immunoglobulin gene expression in transformed lymphoid cells. *Proc. Natl. Acad. Sci. U.S.A.* **80**: 825.

Palmiter, R. D., G. Norstedt, R. E. Gelinas, R. E. Hammer, and R. L. Brinster. 1983. Metallothionein-human GH fusion genes stimulate growth of mice. *Science* **222**: 809.

Pernis, B., G. Chiappino, A. S. Kelus, and P. G. H. Gell. 1965. Cellular localization of immunoglobulins with different allotypic specificities in rabbit lymphoid tissues. *J. Exp. Med.* **122**: 853.

Rice, D. and D. Baltimore. 1982. Regulated expression of an immunoglobulin κ gene introduced into a mouse lymphoid cell line. *Proc. Natl. Acad. Sci. U.S.A.* **79**: 7862.

Ritchie, K. A., R. L. Brinster, and U. Storb. 1984. Allelic exclusion and control of endogenous immunoglobulin gene rearrangement in kappa transgenic mice. *Nature* **312**: 517.

Rokhlin, O. V., M. N. Petrosyan, A. R. Ibaghimov, T. I. Vengerova, and G. T. Bogachova. 1983. Antigenic markers of V domain of mouse MOPC21 IgG_1 L chain. The conformational basis of V_L domain markers and content of MOPC 21 V_L-like immunoglobulins in sera of inbred mouse strains. *Eur. J. Immunol.* **13**: 397.

Sakano, H., K. Hüppi, Heinrich, Günther, and S. Tonegawa. 1979. Sequences at the somatic recombination sites of immunoglobulin light-chain genes. *Nature* **280**: 288.

Seidman, J. G., E. E. Max, and P. Leder. 1979. A κ-immunoglobulin gene is formed by site-specific recombination without further somatic mutation. *Nature* **280**: 370.

Seidman, J. G. and P. Leder. 1980. A mutant immunoglobulin light chain is formed by aberrant DNA- and RNA-splicing events. *Nature* **286**: 779.

Storb, U. and B. Arp. 1983. Methylation patterns of immunoglobulin genes in lymphoid cells: Correlation of expression and differentiation with undermethylation. *Proc. Natl. Acad. Sci. U.S.A.* **80**: 6642.

Storb, U., B. Arp, and R. Wilson. 1980. Myeloma with multiple rearranged immunoglobulin kappa genes: Only one kappa gene codes for kappa chains. *Nucl. Acids Res.* **8**: 4681.

Storb, U., R. Wilson, E. Selsing, and A. Walfield. 1981. Rearranged and germline immunoglobulin κ genes: Different states of DNase I sensitivity of constant κ genes in immunocompetent and nonimmune cells. *Biochemistry* **20**: 990.

Storb, U., K. A. Denis, R. L. Brinster, and O. N. Witte. 1985. Pre-B cells in κ-transgenic mice. *Nature* (in press).

Storb, U., R. L. O'Brien, M. D. McMullen, K. A. Gollahon, and R. L. Brinster. 1984. High expression of cloned immunoglobulin κ gene in transgenic mice is restricted to B lymphocytes. *Nature* **310**: 238.

Wagner, E. F., T. A. Stewart, and B. Mintz. 1981. The human β-globin gene and a functional viral thymidine kinase gene in developing mice. *Proc. Natl. Acad. Sci. U.S.A.* **78**: 5016.

Walfield, A. M., U. Storb, E. Selsing, and H. Zentgraf. 1980. Comparison of different rearranged immunoglobulin kappa genes of a myeloma by electron-microscopy and restriction mapping of cloned DNA: Implications for "allelic exclusion." *Nucl. Acids. Res.* **8**: 4689.

Walfield, A., E. Selsing, B. Arp, and U. Storb. 1981. Misalignment of V and J gene segments resulting in a nonfunctional immunoglobulin gene. *Nucl. Acids. Res.* **9**: 1101.

Wilson, R., J. Miller, and U. Storb. 1979. Rearrangement of immunoglobulin genes. *Biochemistry* **18**: 5013.

Elevated Expression of Human HPRT in CNS of Transgenic Mice

J. TIMOTHY STOUT,* HOWARD Y. CHEN,† C. THOMAS CASKEY,* AND RALPH L. BRINSTER†
*Departments of Medicine and Cell Biology
Baylor College of Medicine
Houston, Texas 77030
†Laboratory of Reproductive Physiology
School of Veterinary Medicine
University of Pennsylvania
Philadelphia, Pennsylvania 19104

OVERVIEW

Fusion genes containing the human hypoxanthine phosphoribosyltransferase (hHPRT) cDNA, the mouse metallothionein gene promoter (MT-I), and the human growth hormone (hGH) polyadenylation signal have been used to establish transgenic mice that express human HPRT. Normal mouse and human HPRT genes are constitutively expressed at low levels in most tissues with expression in regions of the central nervous system elevated fourfold over other tissues (Kelley et al. 1969). Transgenic mice produced functional hHPRT in various tissues; however, high levels of hHPRT were observed in brain tissues of all expressing mice. CNS levels of hHPRT activity correlated with levels of hHPRT mRNA, but not with the number of fusion gene copies. Progeny of these transgenic mice expressed inherited fusion genes at the same level and with the same pattern of tissue distribution as their parents. The fact that mice containing hHPRT fusion genes, integrated at different chromosomal sites and controlled by heterologous promoters, express these genes in a tissue-preferential manner suggests that information within the HPRT cDNA influences tissue variability.

INTRODUCTION

HPRT is a cytoplasmic, multimeric enzyme of the purine salvage pathway that catalyzes the formation of 5′ IMP and 5′ GMP from the bases hypoxanthine and guanine respectively (Krenitsky et al. 1969). Complete deficiency of this enzyme in man results in the Lesch-Nyhan syndrome, a debilitating X-linked neurologic disorder characterized by mental retardation, spasticity, hyperuricemia, and a bizarre tendancy towards self-mutilative behavior (Seegmiller et al. 1967). Patients with partial HPRT deficiency exhibit hyperuricemia and a form of gouty arthritis but display none of the neurologic features that characterize Lesch-Nyhan patients (Kelley and Wyngaarden 1983).

As an enzyme involved in the metabolic salvage of purines, HPRT can be found in all cell types. Purine nucleotide availability depends on de novo purine synthesis as well as the salvage of preformed purine bases through the purine salvage pathway. In normal humans and mice, HPRT levels are highest in tissues of the central nervous system, particularly in regions of the basal ganglia (Hoefnagel 1968). Levels of amidophosphoribosyltransferase (AMPRT), the rate-limiting enzyme in the de novo purine synthetic pathway, are lowest in the basal ganglia (Howard et al. 1970). The relationships between HPRT deficiency, purine nucleotide availability in the brain, and the neurologic features of the Lesch-Nyhan syndrome remain unclear.

In an effort to study expression of the HPRT gene in vivo, we have established transgenic mice that express a human HPRT fusion gene. This approach has proved useful in the examination of *cis*-acting control elements important in gene expression. Recently Swift et al. (1984) have demonstrated that pancreas-specific expression of the rat elastase I gene in transgenic mice is due to regulatory sequences located in the 5′ flanking regions of this gene. Storb et al. (1984) have shown that microinjection of cloned genomic fragments containing a rearranged immunoglobulin κ gene into fertilized mouse eggs will produce mice that express this gene in spleen but not liver tissues. Furthermore, when genomic fragments containing the chicken transferrin gene were incorporated into transgenic mice, McKnight et al. (1983) were able to demonstrate a 5–10-fold preferential expression of this gene in liver tissue, the site of normal expression. These studies indicate that sequences within or nearby genes influence tissue-specific expression in transgenic mice. In this paper we will describe evidence that suggests that information within the HPRT cDNA is influential in increased CNS expression in transgenic mice.

RESULTS

Generation of Transgenic Mice

Figure 1 depicts the fusion gene (pHPT-50) employed in these studies. Human HPRT cDNA was cloned into the *Bgl*II site of an expression vector that provided approximately 770 bp of the mouse metallothionein promoter (a region known to be responsive to Cd^{+2} induction) and 650 bp of 3′ flanking sequence from the human growth hormone gene (a region containing a functional polyadenylation signal [Glanville et al. 1981; Sellburg 1982]). Of 389 mouse embryos, each injected with approximately 900 linearized plasmids, 241 viable embryos were implanted into the uteri of pseudopregnant foster mice (Brinster et al. 1981). There were 45 offspring, 17 of which contained the intact fusion gene stably inserted into the mouse genome. Junctional fragments seen in Southern blots indicated that no two mice had the same genomic integration site (Southern 1975).

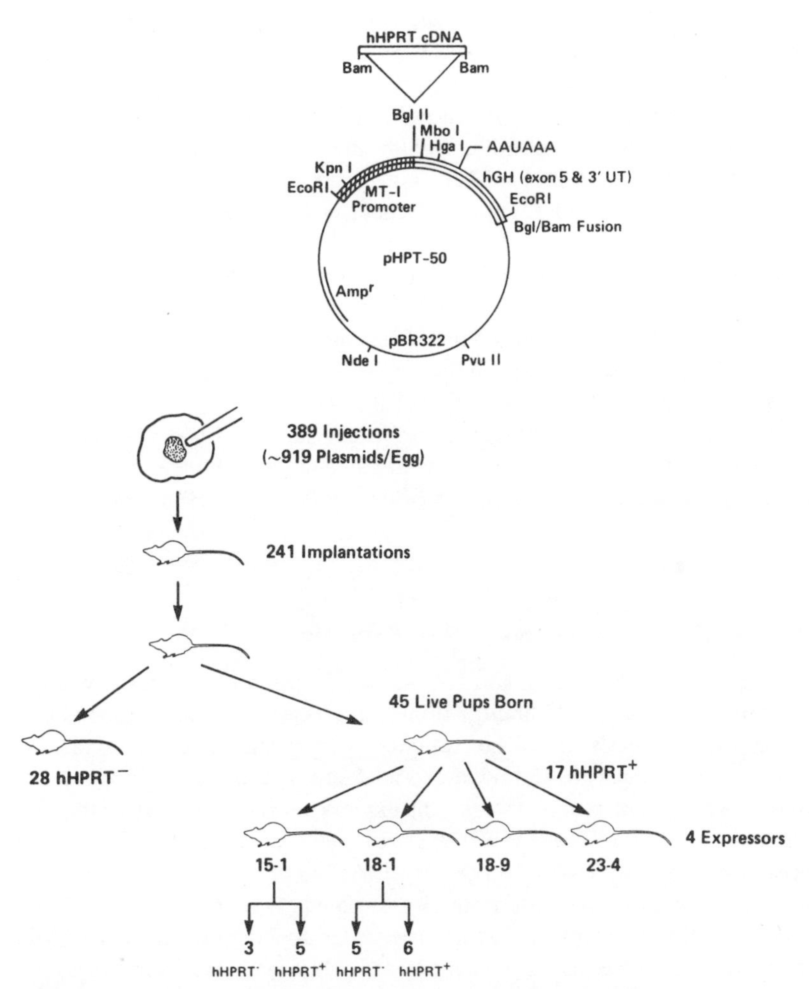

Figure 1
HPRT fusion gene and experimental scheme. Human hHPRT cDNA fragments were ligated to *Bam*HI linkers and fused with the MT-I/hHG expression vector. Male pronuclei of mouse embryos were microinjected with 2 pl containing approximately 900 of these recombinants (Brinster et al. 1981). Embryos were transferred to foster mothers and tail DNA isolated from the resultant pups was used to check fusion gene integration (Southern 1975).

Expression of hHPRT

Expression of the hHPRT gene following Cd^{+2} induction in transgenic mice was examined at both the RNA and protein level. Fractional representation of hHPRT and mHPRT in cell extracts was determined by separation of these proteins on an isoelectric focusing gel followed by an in situ enzyme assay (Chasin and Urlaub

1976; Johnson et al. 1979). Figure 2 shows representative IEF blots from three transgenic mice. Mouse 23-4 and mouse 18-5 produced detectable hHPRT in brain and liver, and brain and kidney, tissues respectively. Mice 18-9 and 15-1 produced hHPRT in all tissues examined. To measure the relative amounts of mHPRT and hHPRT, radioactive spots corresponding to the mouse and human enzymes were cut out and quantitated by liquid scintillation. A direct HPRT enzyme assay was performed on each extract and specific activities for human and mouse enzymes were determined (Olsen and Milman 1974).

Slot blots of RNA from tissues of transgenic mice were used to detect fusion gene transcripts (Fig. 3) (Maniatis et al. 1982). In these studies, 20 μg of total RNA isolated from mouse tissues were immobilized on nitrocellulose and probed with a nick-translated DNA fragment corresponding to the 3′ end of the fusion gene (*Hga*I-*Eco*RI fragment of the 3′ untranslated region of the *hGH* gene).

The presence of fusion gene transcripts correlated exactly with the presence of hHPRT protein in cell extracts from tissues of these animals. Furthermore, there was a correlation between CNS levels of hHPRT mRNA and hHPRT activity in brain tissues.

Quantitation and Distribution of Foreign Gene Expression

It can be seen in Table 1 that expression of the hHPRT gene occurred variably in all tissues with the exception of the brain. All mice expressed the fusion gene at high levels in CNS tissues. Levels of the human enzyme in transgenic mouse brain tissues were elevated as much as 16-fold over endogenous mouse enzyme levels. In addition, in each transgenic mouse, fusion gene expression was consistently elevated in brain tissues over other tissues. No relation between fusion gene copy number and expression at either the RNA or protein level was found.

Fusion gene expression was consistantly associated with an increase in total HPRT activity (mHPRT + hHPRT). Table 2 depicts the fold increase of total HPRT activity in tissues of transgenic mice over average values obtained from nontransgenic animals. Total CNS HPRT activity in these mice was increased up to as much as fourfold over normal values (mouse 23-4). Overproduction of this enzyme in mice had no observable deleterious effect.

Establishment and Characterization of Mouse Lines

Two transgenic male mice (15-1 and 18-5) were mated with nontransgenic females to establish transgenic lines and study fusion gene inheritability. Mouse 15-1, who expressed hHPRT in all tissues, sired eight pups, five of which inherited the human gene. A male from this litter (15-1-1) expressed the human gene in all tissues in a manner quantitatively similar to its parent (Tables 1 and 2). Mouse 18-5, shown to express the fusion gene in brain and kidney tissues, sired 11 pups, six of which re-

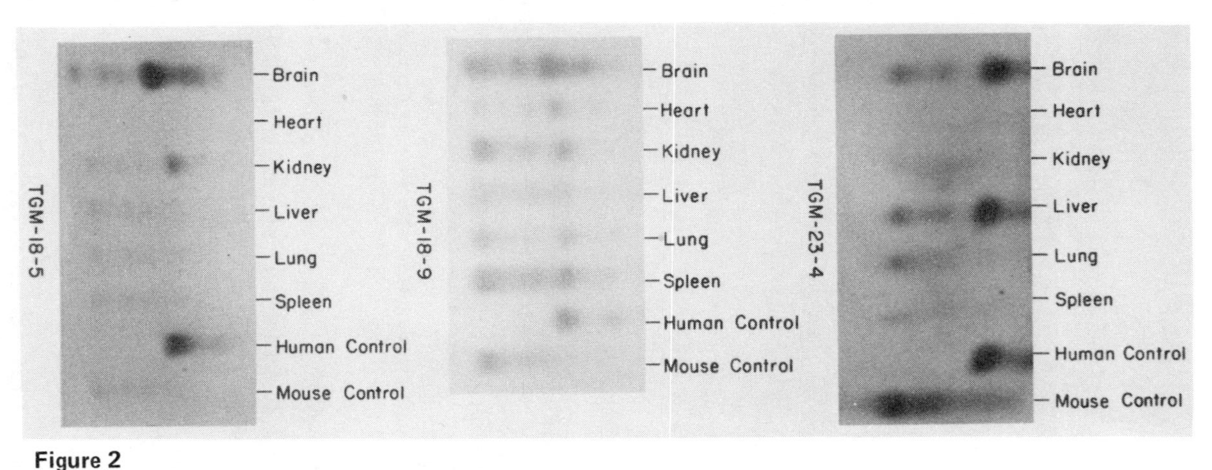

Figure 2
HPRT isoelectric focusing gels. Mice were sacrificed by cervical dislocation 8 hrs following Cd^{+2} treatment (Palmiter et al. 1983). Organs were homogenized in 0.5 ml of 20 mM Tris (pH7.2) and were subsequently subjected to five freeze-thaw cycles. All samples were centrifuged (30,000 × g for 25 min) and 5–10 μl of each supernatent were loaded on nondenaturing isoelectric focusing gels. Samples were focused for 12.5 hr. at 700 v (pH 5–7/21 cm), overlaid with reaction cocktail (100 mM Tris [pH 7.4], 10 mM MgC12, 1μCI [^{14}–C] hypoxanthine [50 μCi/mmole], 10 mM PRPP, 4.8 mg/m^{1} BSA), and incubated at 37°C for 1 hr. A sheet of PEI-cellulose was placed on each gel and [^{14}C]-IMP was allowed to absorb for 15 min at room temperature. Sheets were washed with tap water (3×), dried, sprayed with Enhance, and put against x-ray film for 10–12 hrs at –70°C.

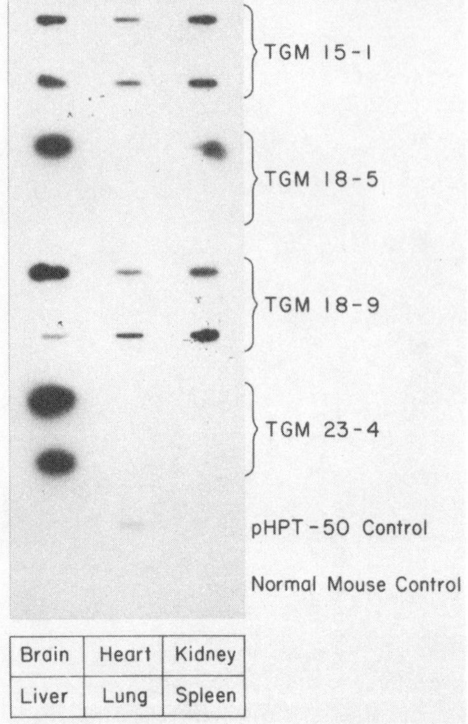

Figure 3

Slot blots of RNA from transgenic mice. Twenty μg of total RNA isolated from transgenic mouse tissues were immobilized on nitrocellulose paper and hybridized to nick-translated *Hga*I-*Eco*RI fragment (sp. act. = 1×10^8 cpm/μg DNA) in 50% formamide, 5 × SSC, 1 × Denhardt's, 0.1% SSC, and denatured herring sperm DNA (250 μg/ml) for 12 hrs at 42°C (Maniatis et al. 1982). RNA from $HPRT^-$ Chinese Hamster cells transfected with pHPT-50 was used as a positive control; RNA from a nontransgenic mouse brain was used as a negative control. Mice and controls were identically treated with Cd^{+2}. The key (bottom) indicates tissue source of RNA for each mouse.

ceived the gene. One of these, mouse 18-5-1 (a male), synthesized hHPRT in brain and kidney tissues only and at levels similar to its parent. Direct male-to-male transmission of this gene with maintenance of tissue-specific expression patterns is indicative of autosomal dominant Mendelian inheritance. Breeding to produce double heterozygotes is progressing.

DISCUSSION

The molecular basis of high MT-hHPRT gene expression in the CNS of Cd^{+2}-stimulated transgenic mice remains unclear. Durnam and Palmiter (1981) have shown that heavy metal induction of genes under the control of the mouse metallo-

Table 1
Tissue Distribution of Human HPRT in Transgenic Mice and Offspring (nmole IMP min^{-1} mg^{-1})

Animal	Tissue					
	Brain	Heart	Kidney	Liver	Lung	Spleen
15-1 (P_1)	8.37	1.92	0.13	0.78	0.57	0.63
15-1-1 (F_1)	11.52	2.54	0.30	1.12	0.68	0.69
18-5 (P_1)	10.91	–	4.85	–	–	–
18-5-1 (F_1)	9.30	–	4.31	–	–	–
18-9	10.48	3.40	2.36	0.99	1.40	4.47
23-4	17.20	–	–	7.93	–	–
Normal control mouse (avg. mHPRT)	4.8	1.2	1.4	1.2	1.4	2.4

Human HPRT activities in each mouse tissue are expressed as a nmole IMP-formed min^{-1} mg of total $protein^{-1}$. Values are given for parental (P_1) and filial (F_1) generations. Normal mouse values are averages compiled from eight nontransgenic control mice.

Table 2
Quantitation of Total HPRT Activity in Transgenic Mice Tissues (nmole IMP min^{-1} mg^{-1})

Animals		Tissue					
		Brain	Heart	Kidney	Liver	Lung	Spleen
15-1	Total activity	8.9	2.1	3.0	3.12	2.84	6.2
	fold increase	1.8	1.7	2.1	2.6	2.0	2.6
15-1-1	Total activity	12.8	3.1	4.1	4.5	4.1	8.6
	fold increase	2.6	2.6	2.9	3.7	3.0	3.6
18-1	Total activity	14.1	–	6.85	–	–	–
	fold increase	3.0	–	4.7	–	–	–
18-1-1	Total activity	13.0	–	6.3	–	–	–
	fold increase	2.7	–	4.5	–	–	–
18-9	Total activity	17.1	5.4	5.9	3.3	3.9	9.8
	fold increase	3.6	4.5	4.2	2.7	2.8	4.1
23-4	Total activity	20.7	–	–	8.6	–	–
	fold increase	4.3	–	–	7.2	–	–
Normal control (avg. mHPRT)		4.8	1.2	1.4	1.2	1.4	2.4

Total HPRT activities (human + mouse) were determined by direct assay and are expressed as nmole IMP-formed min^{-1} mg of total $protein^{-1}$. Values are given for parental (P_1) and filial (F_1) generations. Fold increase over average normal mouse enzyme level for each tissue is also given.

thionein promoter results in an increase of the transcription rate of those genes but that this effect is primarily limited to the liver and kidney. Studies involving the introduction of a metallothionein human growth hormone (MThGH) fusion gene into transgenic mice described low levels of CNS expression of that gene upon Cd^{+2} induction (Palmiter et al. 1983). It appears unlikely that increased CNS expression of the MT-hHPRT gene in mice could be due solely to the presence of the MT promoter or hHG 3′ untranslated sequences. The replacement of growth hormone gene sequences with hHPRT cDNA sequences effects expression of these genes in CNS tissues of mice.

Whether this tissue-specific elevation of hHPRT gene expression occurs via increased hHPRT mRNA synthesis or increased message stability, is unresolved. There are precedents for both mechanisms meditating tissue-specific regulation of genes. Experiments involving MEL cell expression of globin fusion genes have shown that the sequences that stimulate globin gene transcription are located both 5′ and 3′ to the mRNA cap site (Wright et al. 1984; Charnay et al. 1984). Studies by Merrill et al. (1984) have recently shown that the differentiative capacity of the cellular thymidine kinase gene in developing myoblasts is due to sequences that lie between the site of transcription initiation and the 3′ untranslated region. Sequences responsible for the regulation of genes transcribed by RNA polymerase II may be intragenic. Alternatively Jefferson et al. (1984) have demonstrated that the maintenance of high levels of two liver-specific mRNAs did not correlate with the level of transcription of these genes but was apparently due to increased stabilization of these messages. Presumably sequences within these messages effect their selection for stabilization over other transcribed sequences.

This study suggests that sequences within the human HPRT cDNA influence increased expression of a MT-hHPRT fusion gene in the CNS of transgenic mice. High levels of CNS expression of hHPRT occurred without natural promoter or intron sequences. Regional localization of expression within brain tissues of these mice is in progress and may yield information pertinent to the normal regulation of HPRT and the movement and behavioral disorders associated with the Lesch-Nyhan syndrome. Furthermore, this study indicates that increases of total HPRT activity (of at least fourfold) can occur in mouse brain tissue with no obvious consequence. These studies have clear implications for our efforts towards gene replacement therapy for the Lesch-Nyhan syndrome.

ACKNOWLEDGMENT

We thank R. Palmiter for his generous gift of the MThGH expression vector. We also thank Robert Hammer, John Brennand, and Myrna Trumbauer for contributions to this work. This work was supported by NIH grants AM 3142 (C.T.C.) and HD 17321 (R.L.B.). J.T.S. is supported by the Philip M. Berolzheimer Fellowship, and C.T.C. is an Investigator for the Howard Hughes Medical Institute.

REFERENCES

Brinster, R.L., H.Y. Chen, M. Trumbauer, A.W. Senear, R. Warren, and R.D. Palmiter. 1981. Somatic expression of herpes thymidine kinase in mice following injection of a fusion gene into eggs. *Cell* **27**: 223.

Charnay, P., R. Treisman, P. Mellon, M. Chao, R. Axel, and T. Maniatis. 1984. Differences in human α and β-globin gene expression in mouse erythroleukemia cells: The role of intragenic sequences. *Cell* **38**: 251.

Chasin, L. and G. Urlaub. 1976. Mutant alleles for hypoanthine phosphoribosyltransferase: Codominant expression, complementation, and segregation in hybrid chinese hamster cells. *Somatic Cell Genet.* **2**: 453.

Durnam, D.M. and R.D. Palmiter. 1981. Transcriptional regulation of the mouse metallothionein-I gene by heavy metals. *J. Biol. Chem.* **256**: 5712.

Glanville, N., D.M. Durnam, and R.D. Palmiter. 1981. Structure of mouse metallothionein-I gene and its mRNA. *Nature* **292**: 267.

Hoefnagel, D. 1968. Seminars on the Lesch-Nyhan syndrome. Pathology and pathologic physiology. *Fed. Proc.* **27**: 1042.

Howard, W.J., L.A. Kerson, and S.H. Appel. 1970. Synthesis *de novo* of purines in slices of rat brain and liver. *J. Neurochem.* **17**: 121.

Jefferson, D.M., D.F. Clayton, J.E. Darnell, and L.M. Reid. 1984. Posttranscriptional modulation of gene expression in cultured rat hepatocytes. *Mol. Cell. Biol.* **4**: 1929.

Johnson, G.G., L.R. Eisenberg, and B.R. Migeon. 1979. Human and mouse hypoxanthine-guanine phosphoribosyltransferase: Dimers and tetramers. *Science* **203**: 174.

Kelley, W.N. and J.B. Wyngaarden. 1983. Clinical syndromes associated with hypoxanthine phosphoribosyltransferase deficiency. In *The metabolic basis of inherited disease* (ed. J. Stanbury and J.B. Wyngaarden). McGraw-Hill, New York.

Kelley, W.N., M.L. Greene, F.M. Rosenbloom, and J.F. Henderson. 1969. Hypoxanthine-guanine phosphoribosyltransferase in gout. *Ann. Intern. Med.* **70**: 155.

Krenitsky, T.A., R. Papaiannou, and G.B. Elion. 1969. Human hypoxanthine phosphoribosyltransferase. I. Purification, properties and specificity. *J. Biol. Chem.* **244**: 1263.

Maniatis, T., E.F. Fritsch, and J. Sambrook. 1982. *Molecular cloning: A laboratory manual.* Cold Spring Harbor Laboratory, Cold Spring Harbor, New York.

McKnight, S.G., R.E. Hammer, E.A. Kuenzel, and R.L. Brinster. 1983. Expression of the chicken transferrin gene in transgenic mice. *Cell* **34**: 335.

Merrill, G.F., S.D. Hauschka, and S.L. McKnight. 1984. *Tk* enzyme expression in differentiating muscle cells is regulated through an internal segment of the cellular *tk* gene. *Mol. Cell. Biol.* **4**: 1777.

Olsen, A. and G. Milman. 1974. Subunit molecular weight of hypoxanthine-guanine phosphoribosyltransferase. *J. Biol. Chem.* **249**: 4038.

Palmiter, R.D., G. Norstedt, R.E. Gelinas, R.E. Hammer, and R.L. Brinster. 1983. Metallothionein-human GH fusion genes stimulate growth of mice. *Science* **222**: 809.

Seegmiller, J.E., F.M. Rosenbloom, and W.N. Kelley. 1967. An enzyme defect associated with a sex-linked human neurological disorder and excessive purine synthesis. *Science* **155**: 1682.

Sellburg, P.H. 1982. The human growth hormone gene family: Nucleotide sequences show recent divergence and predict a new polypeptide hormone. *DNA* **1**: 239.

Southern, E.M. 1975. Detection of specific sequences among DNA fragments separated by gel electrophoresis. *J. Mol. Biol.* **98**: 503.

Storb, U., R.L. O'Brien, M.D. McMullen, K.A. Gollahon, and R.L. Brinster. 1984. High expression of cloned immunoglobulin κ gene in transgenic mice is restricted to B lymphocytes. *Nature* **310**: 238.

Swift, G.H., R.E. Hammer, R.J. MacDonald, and R.L. Brinster. 1984. Tissue-specific expression of the rat pancreatic elastase 1 gene in transgenic mice. *Cell* **38**: 639.

Wright, S., A. Rosenthal, R. Flavell, and F. Grosveld. 1984. DNA sequences required for regulated expression of β-globin genes in murine erythroleukemia cells. *Cell* **38**: 265.

Gene Transfer into Pronuclei of Cattle and Sheep Zygotes

DUANE KRAEMER, BRIJINDER MINHAS, AND JOE CAPEHART
Department of Veterinary Physiology and Pharmacology
Texas A&M University
College Station, Texas 77843

OVERVIEW

Research is being conducted in an effort to develop the capability for gene transfer in cattle and sheep. The approach is patterned after the pronuclear injection procedure that has been used by others to transfer genes in mice. A DNA stain (4′-6′-diamidino-2-phenylindole [DAPI]) is used to induce fluorescence of the pronuclei when exposed to ultraviolet light since the pronuclei of cattle zygotes are masked by yolk material in the vitellus. An average of 3.1 zygotes were obtained per superovulated bovine donor. Herpes simplex virus thymidine kinase (HSV-TK) genes were microinjected into the male pronuclei of 85 bovine zygotes. Growth of stained, gene-injected zygotes was obtained both in vitro and in vivo. Forty-one such zygotes were transferred to the oviducts of 11 synchronized recipient bovine females. Two bovine pregnancies have continued to term, producing three male offspring. Data are not yet available regarding incorporation of the HSV-TK gene into the bovine genome. A single male ovine fetus was aborted by a recipient ewe 117 days after the transfer of four gene-injected zygotes to her oviducts.

These data show that bovine zygotes can survive DAPI staining and gene injection into their male pronucleus and that the resulting offspring appear to be normal.

INTRODUCTION

Variability is one of the major factors influencing the efficiency of selective breeding of livestock. The use of recombinant DNA technology offers the potential for greatly increasing the genetic variability within a species or strain of animal. Although the potential benefits from gene transfer for increasing the efficiency of food and fiber production are theoretically great, definition of the true value of such techniques awaits experimentation.

We have begun the process of exploring the possibilities of gene transfer in cattle and sheep. Our approach has been to adapt techniques which have been used by others to transfer foreign genes in the mouse, for use in these species of livestock. Details of this work have been published in a Ph.D. Dissertation (Minhas 1984).

RESULTS

Crossbred donor heifers were superovulated by administration of 5 mg porcine follicle stimulating hormone (FSH) twice daily for 5 days beginning on day 9 or 10 of the estrous cycle. Prostaglandin $F_{2\alpha}$ (30 mg) was administered at the fifth injection of FSH and the cows were either artificially inseminated or naturally mated at estrus. The oviducts were flushed surgically to obtain the embryos 48 hours after the beginning of estrus. A total of 279 ovulations were observed and 207 ova (74%) were obtained from the 30 superovulated heifers. Of the 207 ova collected, 160 (77%) were at the one-cell stage, 37 (18%) were at the two-cell stage, and 10 (5%) were at the four-cell stage of development (Table 1). Fifty-eight percent (58%) of the one-cell ova were fertilized, thereby yielding 3.1 zygotes per donor (Table 1). The one-cell ova were stained with 4′-6′-diamidino-2-phenylindole (DAPI) and observed by fluorescence microscopy (Fig. 1) to determine whether or not they were fertilized since the pronuclei of bovine zygotes cannot be visualized by incandescent light microscopy (Minhas et al. 1984). The herpes simplex virus thymidine kinase (HSV-TK) gene in the circular PBR 322 plasmid was microinjected into 85 of the 92 zygotes observed.

Fifteen of the gene-injected zygotes were cultured in Ham F-10 medium supplemented with 20% heat inactivated fetal calf serum and 100 IU/ml penicillin, 100 μg/ml streptomycin, and 0.25 μg/ml fungizone in a 5% CO_2 in air atmosphere at 37°. Eight of the 15 embryos (40%) developed in culture; three to the two-cell stage, two to the four-cell stage, and one to the early morula stage of development.

Twenty-three of the gene-injected zygotes were surgically transferred to the oviducts of three recipients which showed no signs of ovulation nor pending ovulation. The reproductive tracts were surgically flushed at 48–72 hours after transfer. The stages of development of the nine embryos which were recovered (39%) are shown in Table 2. Seven of the nine (78%) embryos had developed during in vivo culture. One six-cell embryo was transferred surgically to the oviduct ipsilateral to the corpus luteum of a synchronized recipient, but the pregnancy did not continue. An eight-cell and a four-cell embryo were transferred nonsurgically to the ipsilateral and contralateral uterine horns, respectively, of a day 4 recipient, but the pregnancy did not continue.

Table 1
Stage of Development of Cattle Ova When Collected 48 Hours After Onset of Estrus

Stage	Number obtained	Percentage
One-celled ova	160	160/207 = 77.3%
Zygotes	92	92/160 = 57.5%
Two-celled embryos	37	37/207 = 17.9%
Four-celled embryos	10	10/207 = 4.8%

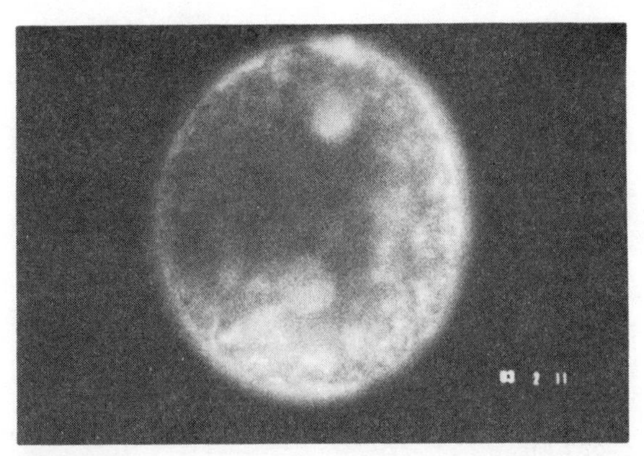

Figure 1
A bovine zygote which has been stained by DAPI to reveal the two pronuclei

Forty-one of the gene-injected zygotes were transferred to the oviducts of 11 synchronized bovine recipients. The distribution of the transferred zygotes between the recipients and the pregnancy outcomes are shown in Table 3. The number of zygotes transferred per recipient varied from two to five, depending upon the number of injected zygotes available per experiment. Two of the 11 recipients (18%) continued their pregnancies to term. One recipient which had received four gene-injected zygotes, two per oviduct, gave birth to a single male offspring (Fig. 2). Blood typing data rule out the recipient as the genetic mother of this offspring and are consistent with blood types of the donor and sire. A second recipient, which received five gene-injected zygotes, gave birth to two male offspring (Fig. 3). Blood typing data are not available on these offspring at this time,

Table 2
Embryological Development of Gene-injected Bovine Zygotes Following In Vivo Incubation

Recipient number	Ovary condition	Number of zygotes per recipient	Duration of in vivo culture	Embryos recovered
7000	No ovulation	7	48 hr	1 zygote 2 two-celled 1 six-celled
7524	No ovulation	13	72 hr	1 zygote 1 four-celled 1 six-celled
7201	No ovulation	3	60 hr	1 four-celled 1 eight-celled

Table 3
Numbers of Gene-injected Bovine Zygotes Transferred per Synchronized Recipient and the Pregnancy Results

Number of recipients	Corpus luteum Left	Corpus luteum Right	Zygotes per recipient	Number pregnant	Number of offspring
2	1	1	5 (3 ipsilateral) (2 contralateral)	1	2
6	1	5	4 (2 to each oviduct)	1	1
1	1	–	3 (2 ipsilateral) (1 contralateral)	0	0
2	1	1	2 (1 to each oviduct)	0	0

Figure 2
The first bovine offspring produced following DAPI staining and microinjection of the HSV-TK gene into the male pronucleus

Figure 3
Two male bovine offspring carried by the same recipient following the transfer of five DAPI-stained, HSV-TK gene-injected zygotes

nor have the tests for incorporation of the HSV-TK gene into the genome of the three offspring been completed to date.

One attempt was made to transfer the HSV-TK gene to the pronuclei of sheep zygotes. The donor was superovulated by administering 14 daily injections of 15 mg progesterone plus 1000 IU PMSG on the last day of progesterone administration. The donor was mated naturally at estrus and the zygotes were collected surgically 30 hours after the beginning of estrus. The gene was injected into the male pronuclei of two zygotes using the DAPI stain to visualize the pronuclei, and in two the gene was injected using phase contrast microscopy to visualize the pronuclei near the periphery of the cell. All four of the injected zygotes were transferred to a single synchronized recipient, placing the two DAPI-stained zygotes into one oviduct and the other two into the other oviduct. This recipient remained pregnant until a single male fetus was aborted at 117 days of gestation. Examination of the uterus after the abortion revealed that the pregnancy was on the side to which the nonstained zygotes had been transferred.

DISCUSSION

These data show that bovine zygotes can survive the combined procedures of staining the DNA with the fluorescent stain, DAPI, and the microinjection of foreign DNA (HSV-TK gene) into the male pronucleus. The three offspring produced appear to be normal in spite of the potential mutagenicity of the ultraviolet irradiation used to visualize the pronuclei. Bovine pronuclei cannot be visualized by phase contrast or interference microscopy as can the pronuclei of mice, rabbits, and primates. The yolk material in the bovine zygote, and to a lesser extent in the sheep zygote, masks the pronuclei, thereby requiring the development of special procedures for observing the pronuclei during microinjection. Another DNA stain, Hoescht 33342, has been used by Critser et al. (1983) for visualizing pronuclei and nuclei in living bovine and mouse embryos. Centrifuging the zygotes to shift the yolk material to one side of the vitellus has been used to visualize pronuclei in porcine zygotes (Wall et al. 1984). Nancarrow et al. (1984) found that although centrifugation did not significantly affect the viability of sheep embryos, it did not aid in visualizing the nuclei. There are no reports to date of the use of centrifugation for visualization of pronuclei in bovine zygotes.

The application of gene transfer for genetic improvement of livestock via the pronuclei requires the availability of relatively large numbers of zygotes. We found that approximately 48 hours after the beginning of estrus in cattle and 30 hours after the beginning of estrus in sheep are appropriate times to obtain zygotes from the oviducts of each of these species. However, the surgical procedures used to recover the zygotes from the oviducts induce adhesions between the ovaries and oviducts, thereby limiting the number of times that zygotes can be collected from an individual female. This limitation would be undesirable in the case of genetically valuable females which would be used in commercial application of gene transfer for livestock improvement. Therefore, it will be important either to develop laparoscopic procedures for collection of zygotes from the oviducts or to develop efficient methods for in vitro fertilization of ova collected via laparoscopy from the ovaries of the livestock species.

ACKNOWLEDGMENTS

The authors express appreciation to Dr. M. J. Bowen for assistance with the surgery involved in this research, Drs. T.E. Wagner and J.E. Womack for supplying the HSV-TK gene, Dana Williamson for technical assistance, numerous undergraduate and graduate students who assisted with various aspects of animal care and surgery, and Tammy Cooper for assistance in the preparation of this manuscript. This research was supported by The Hillcrest Foundation, Dallas, Texas; Dessert Seed Co., Inc., Elcentro, California; The Robert J. Kleberg, Jr. and Helen C. Kleberg Foundation, San Antonio, Texas; and the Texas Agricultural Experiment Station.

REFERENCES

Critser, E.S., G.D. Ball, M.J. O'Brien, and N.L. First. 1983. Use of a fluorescent stain for visualization of nuclear material in living gametes and early embryos. *Biol. Reprod.* (Suppl. 1) **28**: 140.

Minhas, B.S. 1984. Gene transfer to mammalian zygotes. Ph.D. dissertation. Texas A&M University, College Station, Texas.

Minhas, B.S., J.S. Capehart, M.J. Bowen, J.E. Womack, J.D. McCrady, P.G. Harms, T.E. Wagner, and D.C. Kraemer. 1984. Visualization of pronuclei in living bovine zygotes. *Biol. Reprod.* **30**: 687.

Nancarrow, C.D., J.D. Murray, M.P. Boland, I.G. Hazelton, and R. Sutton. 1984. Towards gene transfer into ruminant embryos: Effect of centrifugation. *Theriogenology* **21**: 248.

Wall, R.J., V.G. Pursel, and R.E. Hammer. 1984. Normal development of porcine embryos after centrifugation. *J. Anim. Sci.* (Suppl. 1) **59**: 333.

Session 4: Nonmammalian Systems

A Sea Urchin Gene Transfer System

KAREN S. KATULA, BARBARA R. HOUGH-EVANS, ANDREW P. McMAHON, CONSTANTIN N. FLYTZANIS, ROBERTA R. FRANKS, ROY J. BRITTEN, AND ERIC H. DAVIDSON
Division of Biology
California Institute of Technology
Pasadena, California 91125

OVERVIEW

We are interested in studying the regulation of gene expression during development of the sea urchin *Strongylocentrotus purpuratus*. A method has been developed to introduce DNA into the cytoplasm of the unfertilized egg by microinjection. We have examined the fate of the injected DNA throughout the life cycle of the sea urchin and its expression in early embryogenesis. Cloned plasmid DNA injected in a linear form was found to ligate and become amplified during the early stages of embryogenesis. The exogenous DNA persisted throughout larval life and was detected in juvenile sea urchins at 2 months postmetamorphosis. Twelve sexually mature sea urchins grown from injected eggs were analyzed for the presence of exogenous DNA in their gametes. In one case, cloned plasmid DNA was detected in the sperm from a male. To test for expression, a plasmid containing the regulatory sequences for the *Drosophila* 70-kd heat shock protein fused to the bacterial enzyme, chloramphenicol acetyltransferase (CAT), was injected into unfertilized eggs. Heat inducible CAT activity was detected in extracts from injected 3-day-old larvae. These results indicate the potential for investigating gene regulation in the sea urchin by reintroducing DNA into the developing organism.

Introduction

Developmental biologists have for many years attempted to understand the mechanisms by which genes are regulated in a temporal and tissue-specific manner during embryogenesis. These attempts have led recently to the development of methods of introducing genes of interest into developing organisms. By modifying the introduced gene it should be possible to identify the nucleic acid sequences required for regulated expression. Significant progress in generating germline transformants has been achieved in *Drosophila* by use of P-factor vectors and in the mouse by pronuclear injection. A number of developmentally regulated genes have been shown to be correctly expressed in *Drosophila* (Goldberg et al. 1983; Scholnick et al. 1983; Spradling and Rubin 1983) and in the mouse (this volume).

The developing sea urchin offers advantages that makes it particularly suitable as a model system for studying the molecular basis of early development. More is probably known concerning the embryology and molecular biology of the sea urchin than for any other organism (Davidson 1976; Davidson et al. 1982; Angerer and Davidson 1984). The female sea urchin produces large numbers of eggs that when fertilized develop rapidly and synchronously, making feasible biochemical investigations of the early stages of embryogenesis. The lineage relationships among the early embryonic cells leading to primary tissue differentiation are well worked out; and, in addition, the methodology is available to fractionate some of the cell types. The transparency of the embryo and its relatively small numbers of cells have made it possible to identify a variety of lineage-specific genes using *in situ* hybridization with radioactive nucleic acid probes and immunofluorescence staining (McClay et al. 1983; Angerer and Davidson 1984; Carpenter et al. 1984; K. H. Cox et al., in prep.). These cloned genes should enable us to investigate a number of questions concerning ontogenic gene regulation.

We have developed a simple and efficient means of introducing DNA into the cytoplasm of the unfertilized sea urchin egg using the technique of microinjection (McMahon et al. 1985). Briefly, unfertilized eggs are dejellied in low pH sea water and then fixed in rows onto protamine sulfate-treated dishes. Approximately 2 pL of a DNA solution is injected into each egg from a continuously flowing glass needle. Using this protocol we have found that 90-95% of the injected eggs fertilize, and of these 40-50% reach the pluteus stage of development (72-96 hours). With some practice a person can inject 200-300 eggs in one hour.

RESULTS

Fate of the Injected DNA

We have examined the fate of the injected DNA at various times in the life cycle of the sea urchin, including the embryo, larva, juvenile, and finally in the germline of the sexually mature sea urchin (Flytzanis et al. 1985; McMahon et al. 1985). When DNA was extracted from embryos 24-72 hours after injection and displayed by electrophoresis on agarose gels, the injected sequences were found in a high molecular weight form. Figure 1 shows an example of such an experiment, in which about 9000 molecules of a linearized plasmid DNA were injected per egg. After the extracted DNA was digested with a different restriction enzyme than had been used to linearize the plasmid, two prominent bands of 8.8 kb and 5.1 kb were observed after hybridization. These fragments correspond to the expected lengths if the linearized plasmids ligate after injection in a random manner as diagrammed in Figure 1b.

The amount of DNA in the bands seen in Figure 1 and similar experiments was quantitated by comparison to standard amounts of DNA, using scintillation counting or densitometry (data not shown). It was clear from these data that the exogenous DNA has been extensively amplified and is present in much greater amounts than originally injected. The average net amplification of exogenous DNA was 25-fold; and in many recent experiments, levels of 100-fold have been obtained. The ligation and subsequent amplification of the exogenous DNA requires

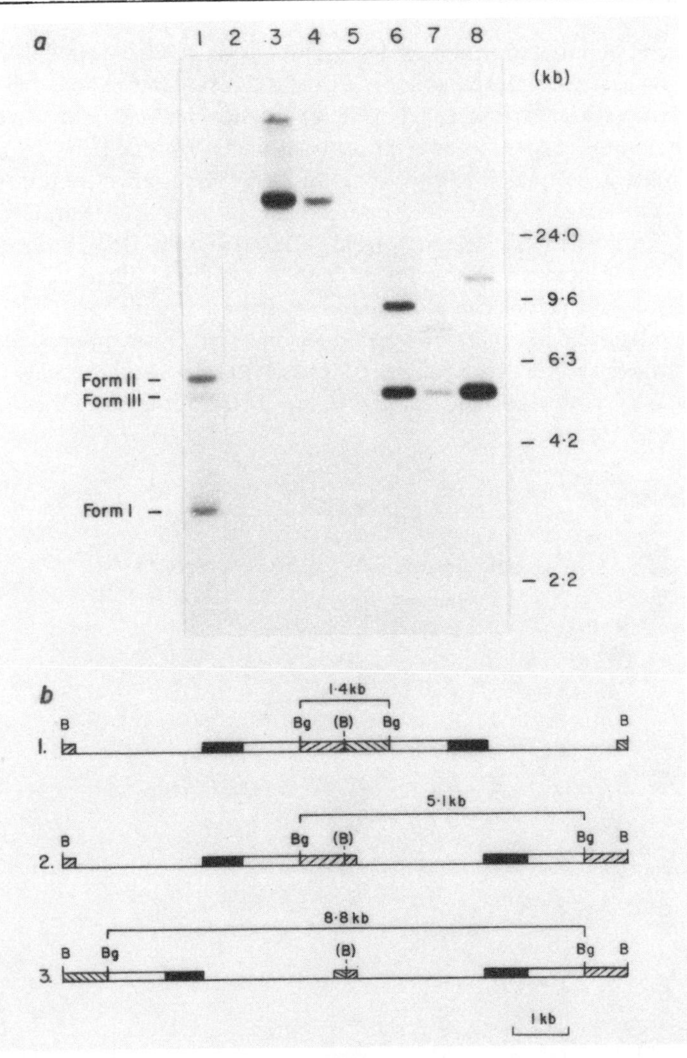

Figure 1
Fate of linearized plasmid DNA injected into the sea urchin egg. (*a*) About 9000 copies of the plasmid pISA linearized with *Bam*HI or *Hinc*II were injected into unfertilized eggs. pISA contains 5′ flanking sequence of the *S. purpuratus* C1 actin gene (Lee et al. 1984) ligated to the TN5 aminoglycoside 3′ phosphotransferase gene (Davies and Smith 1978), which is flanked by the 3′ poly(A) addition site from the herpes virus thymidine kinase gene (McKnight 1980). (Lane *1*) 3×10^6 Molecules of pISA in the three indicated conformational states; (lane *2*) undigested DNA from 64 undigested gastrulae; (lane *3*) undigested DNA from 67 gastrulae injected with *Bam*HI-linearized pISA; (lane *4*) undigested DNA from 64 gastrulae injected with *Hinc*II-linearized pISA; (lane *5*) *Bgl*II digestion of DNA from 55 uninjected gastrulae; (lane *6*) *Bgl*II digestion of DNA from 55 gastrulae injected with *Bam*HI-linearized pISA; (lane *7*) *Bgl*II digestion of DNA from 55 gastrulae injected with *Hinc*II-linearized pISA; (lane *8*) *Bam*HI digestion of DNA from 55 gastrulae injected with *Bam*HI-linearized pISA. (*b*) Concatenated forms of pISA that would arise following random end-to-end ligation of linearized molecules at the *Bam*HI sites. The band corresponding to the 1.4-kb fragment that would be generated if the plasmids ligated as shown in the first possibility is not visible since it lacks homology to the probe used in this experiment. This band does appear if the whole plasmid is used as a probe. Brackets indicate the size of the predicted fragments that would be released upon *Bgl*II digestion. Reprinted, with permission, from McMahon et al. (1985).

that the plasmid be injected in a linear form. If the plasmid was injected in a supercoiled form, no ligation of the exogenous DNA was observed (McMahon et al. 1985); and if linearized with a restriction enzyme that generates blunt rather than overhanging ends, the amount of ligation and amplification is much decreased (Fig. 1a, lane *7*).

Following the initial period of embryogenesis, from fertilization to formation of the pluteus, the larvae begin to feed and increase in mass. To determine if the exogenous DNA persists through the period of larval growth, DNA was extracted from 5-week-old feeding larvae and dotted onto nitrocellulose filters. These filters were hybridized with radioactive probes representing the injected plasmids. Figure 2 displays results from a number of experiments in which diverse plasmids were injected. From these and similar experiments, we estimate that on the average 55% of the larvae raised from eggs injected with linear DNAs retain the exogenous se-

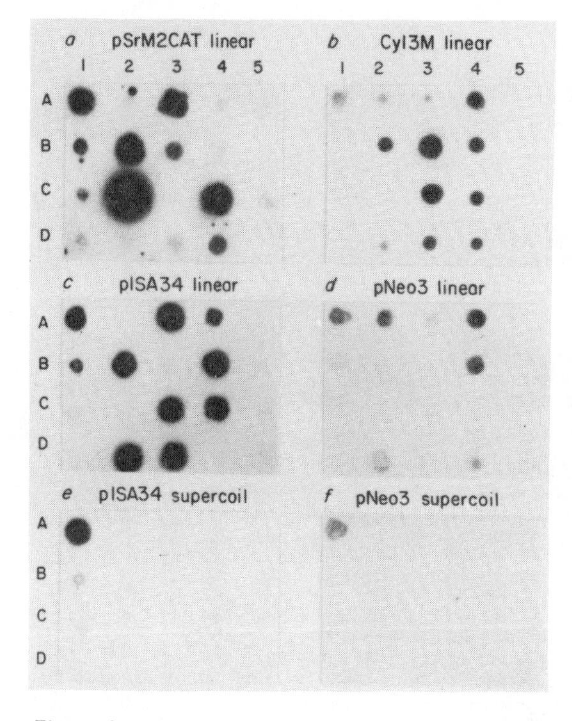

Figure 2
Persistence of exogenous DNA sequences in 5-week cultured larvae. Plasmid DNA was injected in either a supercoiled or linear form into the cytoplasm of unfertilized eggs. At 5 weeks postfertilization, DNA was extracted from individual larvae and blotted onto nitrocellulose as described (Flytzanis et al. 1985). The filters were hybridized to probes homologous to the injected DNA. Column 1 contains standard quantities of the injected plasmids: (*A*) 1.25×10^6 Molecules; (*B*) 2.5×10^5 molecules; (*C*) 5×10^4 molecules; (*D*) 1×10^4 molecules. Lanes *2*, *3*, and *4* contain DNA from individual injected larvae, and lane *5* is DNA from uninjected larvae. For a description of the plasmids see Flytzanis et al. (1985), from which these data are reproduced. Reprinted, with permission, from Flytzanis et al. (1985).

quences. Consistent with the results obtained in early embryos, no exogenous DNA sequences are detected when the plasmid is initially injected in supercoiled configuration (Fig. 2e and f). No significant differences in the extent of amplification have been observed among the different plasmids injected. Since these plasmids contain a variety of diverse sequences it appears that the amplification process does not require the presence of a particular DNA sequence or of eukaryotic DNA.

Sea urchin larvae undergo metamorphosis 5–6 weeks after fertilization. During metamorphosis most of the larval tissues are destroyed and the juvenile sea urchin arises from the imaginal cells of the rudiment. If the exogenous DNA is to be carried through to the adult, it is necessary that this DNA be present in the imaginal cells. DNA extracted from individual 2- to 3-month-old juveniles was digested with an appropriate restriction enzyme and subjected to DNA blot analysis. DNA of 58 individual juveniles from three groups of injected eggs was analyzed. The number of juveniles containing exogenous sequences ranged from 5% to 15% (Flytzanis et al., 1985). Comparing these values to the 55% positives observed in 5-week larvae it appears that not all the cells in many larvae contain exogenous DNA.

A more thorough analysis was made of the DNA extracted from one of the juvenile urchins containing exogenous DNA to determine whether this DNA was integrated into the genome. A genome blot of DNA from this individual is shown in Figure 3a. In addition to the expected bands resulting from a random end-to-end concatenate of the linear molecules originally injected (heavy arrows), there are a number of additional fainter bands (light arrows). These bands might result from junction points between the exogenous DNA and sea urchin genomic DNA, or else represent deletions or recombinations within the concatenate. To determine the actual organization of this DNA, an aliquot was used to construct a recombinant DNA library in the λ cloning vector EMBL3. By screening the library with a combination of probes, it was possible to isolate potential junction fragments. These clones were analyzed for the presence of sea urchin DNA. Two such isolated fragments containing both exogenous sequences and sea urchin DNA are diagrammed in Figure 3b. The existence of a junction between the injected DNA and sea urchin genomic DNA and the exact site of integration were demonstrated by nucleotide sequencing (Flytzanis et al. 1985).

Finally, we wanted to determine if any integrated, exogenous DNA sequences could be detected in the germline of animals which developed from injected eggs. Gametes were obtained from 12 mature sea urchins grown from eggs injected with plasmids containing *Drosophila* P-elements. As stated above, all of our evidence suggests that the same results occur irrespective of the sequence injected. DNA was extracted directly from the sperm or from 3-day-old larvae derived from eggs fertilized with sperm from a noninjected sea urchin. The isolated DNA was digested, blotted, and hybridized with probes homologous to the exogenous DNA. Of these 12 animals one was found to contain exogenous sequences in the DNA extracted from its sperm. A genome blot of this DNA is shown in Figure 4.

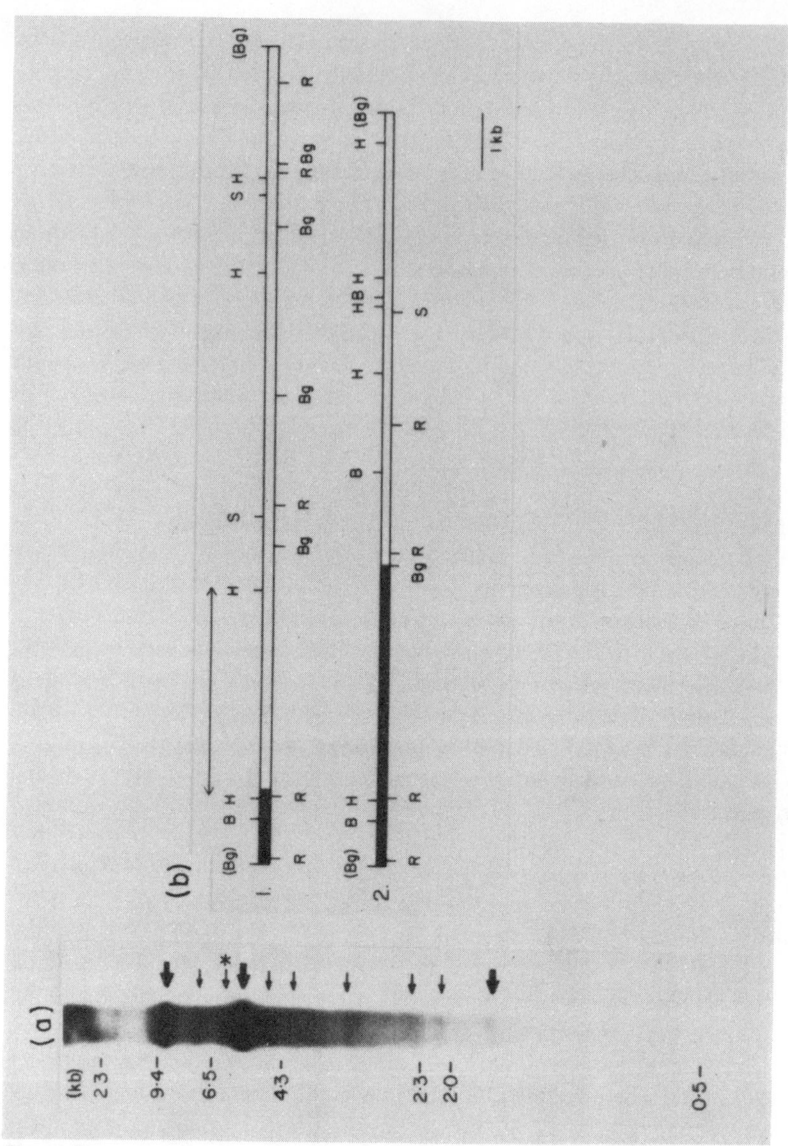

Figure 3
Integration of exogenous DNA in genomic DNA from a juvenile sea urchin. (*A*) DNA was extracted from a 2-month-old juvenile grown from an egg injected with 9000 molecules of the linearized plasmid pNeo3. pNeo3 contains the TN5 amino glycoside 3′ phosphotransferase gene linked to a promoter and poly(A) addition site from the herpes virus thymidine kinase gene cloned into pBR322. The DNA was electrophoresed either undigested or after digestion with *Bgl*II and blot hybridized with a probe to pNeo3. Heavy arrows indicate the bands expected to be generated on *Bgl*II digestion of a random end-to-end concatenate of the pNeo3 plasmid. Other light bands are marked by fine arrows. (*b*) Cloned DNA fragments from an EMBL3 genomic library made with the DNA extracted from the individual in *a*. The open bars represent sea urchin DNA and the closed bar pNeo3 sequences. Reprinted, with permission, from Flytzanis et al. (1985).

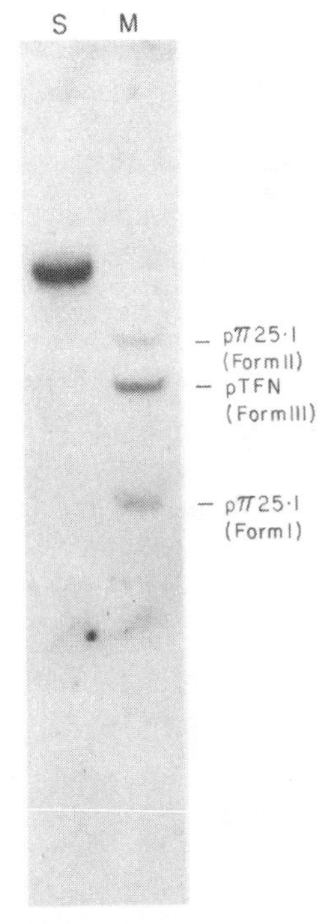

Figure 4
Presence of exogenous DNA sequences in sperm DNA. Genome blot of DNA obtained from sperm of a mature sea urchin grown from an egg injected with the plasmids pπ25.1 (Spradling and Rubin 1982) and pTFN (Flytzanis et al. 1985). The sperm DNA was digested with the restriction enzyme, *Bgl*II, and electrophoresed along with the purified plasmids digested with the same enzyme. *Bgl*II linearizes pTFN but does not cut pπ25.1 (the two bands corresponding to pπ25.1 resulted from some of the supercoiled plasmid's becoming nicked). The hybridization probes were the complete plasmids, nick translated. (*S*) Sperm DNA; (*M*) plasmid marker DNA.

Expression of Injected DNA

To test for expression of injected DNA sequences, we initially used the plasmid hsp-CAT1 (DiNocera and Dawid 1983), which contains the bacterial structural gene for CAT, fused to the 5′ regulatory region of the *Drosophila* 70-kd heat shock protein gene. For analysis of protein synthesis in *S. purpuratus* embryos after heat shock, we chose 25°C, based on data obtained by B. Brandhorst (pers. comm.), and additional experiments carried out by ourselves (McMahon et al. 1984). Embryos of *Stronglylocentrotus purpuratus* are raised in the laboratory at 15°C. Pluteus larvae were exposed to 25°C for 2, 3, and 5 hours and the newly synthesized proteins labeled by adding [^{35}S]methionine during the last hour of incubation. A general reduction in the amount of protein synthesis was observed following heat shock, along with the appearance of several new proteins. This response is similar to what has been observed in other heat shocked systems. Since the change in protein synthesis was maximal within 2 hours, this time was chosen to assay for expression of the hsp-CAT1 plasmid. Approximately 7500 molecules of the *Bgl*II-linearized plasmid were injected, and at 72 hours postfertilization, the pluteus larvae were analyzed for the presence of exogenous DNA and heat inducible CAT activity. The linearized hsp-CAT1 plasmid was found to have been ligated and replicated during early embryogenesis, as expected from our earlier results. By pluteus stage the exogenous DNA had amplified about 100-fold. CAT activity was assayed after exposure of the plutei to 25°C for 2 hours (Fig. 5). No CAT activity was observed in the uninjected controls, whereas a low level of activity was observed in the embryos injected with hsp-CAT1 but maintained at 15°C. Following heat shock, a great increase in CAT activity was observed, indicating that it is heat inducible. The amount of CAT activity was measured as 8–10 times greater than in the injected but unshocked larvae, and it results from the synthesis of about one million molecules of CAT enzyme proteins per embryo during the 2 hours of heat treatment (McMahon et al. 1984).

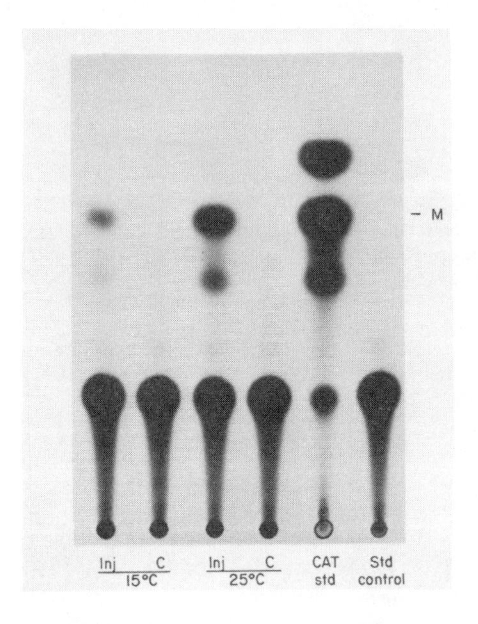

Figure 5
CAT enzyme activity in embryos containing the plasmid, hsp-CAT1. Injected (*Inj*) and control (*C*) eggs were grown to the pluteus stage and exposed to either normal culture temperatures (15°C) or heat shock conditions (25°C) for 2 hours prior to assaying for CAT activity (for details of sample preparation and enzyme assay, see McMahon et al. 1984). As controls for the assay, commercial bacterial CAT enzyme (*CAT std*) or no enzyme or extract (*std control*) was added to the reaction mixture. (*M*) indicates the predominate reaction product, the monoacetylated form of chloramphenicol. Reprinted, with permission, from McMahon et al. (1984).

DISCUSSION

The object of our efforts to develop a gene transfer system in the sea urchin is to investigate the regulation of various sea urchin genes active during early embryogenesis. If it is to be useful for answering questions dealing with ontogenic regulation, this experimental approach requires that the introduced DNA persist and be expressed similarly to the endogenous genes. We have shown that when injected into the cytoplasm of the unfertilized egg, linearized DNA becomes rapidly ligated and is amplified during early embryogenesis. This DNA was found to persist in the majority of the sea urchins throughout larval life, and in a smaller number, in the postmetamorphosed juveniles as well. The discovery of an adult containing exogenous DNA in its germline indicates it should be possible to accumulate, in time, a group of individuals that have been transformed. From each such animal we can anticipate an ample supply of gametes containing the exogenous genes for years to come, as at least in laboratory culture, this species remains reproductively active over very long periods (Leahy et al. 1978).

The relatively small number of postmetamorphosis animals containing exogenous DNA sequences indicates that many of the injected larvae are probably mosaic. Such mosaicism would result if the DNA becomes integrated at different times during development, and in only some of the embryonic cells. In only those cases where the DNA has integrated in cells giving rise to the adult will DNA be found in the juvenile and possibly the germline. One potential means of decreasing the incidence of mosaic recipients would be to inject the DNA directly into the egg pronucleus or zygotic nucleus. We have recently developed the technology for accomplishing this as well and are now comparing the results obtained with nuclear versus cytoplasmic DNA introduction.

In addition to generating adult sea urchins containing germline integrations of exogenous DNA, we are now investigating the expression of introduced DNA sequences in embryos derived from injected eggs. The feasibility of such a short-term expression system is supported by the results obtained with the hsp-CAT1 plasmid. It is clear that exogenous DNA injected into the cytoplasm of unfertilized sea urchin eggs can be inducibly expressed. Apparently, at least some of the injected DNA is in a physical conformation and cellular location that permits functional interactions with both transcription factors and regulatory proteins. It is particularly interesting that heat inducibility is observed at 25°C, which is the normal culture condition for *Drosophila*. Presumably, the production of a *trans*-acting regulatory protein is stimulated in the sea urchin at 25°C, which is capable of interacting with conserved 5′ sequences in the *Drosophila* heat shock gene. The hsp-CAT fusion gene has also been shown to be inducible under host cell stress conditions in *Xenopus* oocyte nucleus (Voellmy and Rungger 1982), mouse cells (Corces et al. 1981), and yeast cells (Lis et al. 1982).

We cannot say at this time whether injected sea urchin genes will be expressed and, if so, in an ontogenic manner. There is the possibility that ontogenic regulation will take place only if the exogenous DNA is integrated and has been carried through the complete developmental history of the animal. We are presently testing for expression following injection of the various sea urchin actin genes whose 5′ sequences have been fused to the CAT enzyme. One of the most interesting aspects of these experiments is the potential to study specific cell lineage expression. The various sea urchin actin genes have been shown to be active in only particular cell lineages (Angerer and Davidson 1984). By using an antibody to the CAT enzyme, it will be possible to determine if these actin genes are expressed both at the appropriate developmental time and in the proper cell lineage. This would open the way to experimental dissection of the genomic sequences required for the initial activation of these markers of early embryonic cell differentiation.

ACKNOWLEDGMENTS

This research was supported by NIH Grant HD-05753. K.S.K. and R.R.F. were supported by an NIH Postdoctoral Training Grant (HD-07257); C.N.F. by an American Cancer Society, California Division, Lievre Fellowship (S-11-83); and A.P.M. by the same Society (J-33-82).

Note Added in Proof

Since preparation of this article, further experiments carried out in our laboratory have clarified two issues raised in the Discussion. First, we now know that injection of exogenous DNA into the zygote nucleus does indeed increase the frequency with which this DNA can be recovered from the genomes of postmetamorphic juvenile sea urchins (R. R. Franks, unpubl.). Second, we have found that sea urchin genes normally active in an ontogenetically specific manner in the embryo are regulated correctly when introduced into the egg cytoplasm as described. This has been demonstrated for the CyIIIa actin gene and the αH2a histone gene (C. N. Flytzanis, unpubl.), for the CyI actin gene (K. S. Katula, unpubl.), and for the CyIIa actin gene (R. R. Franks, unpubl.).

REFERENCES

Angerer, R. C. and E. H. Davidson. 1984. Molecular indices of cell lineage specification in the sea urchin embryo. *Science* **226**: 1153.

Carpenter, C. D., A. M. Bruskin, P. E. Hardin, M. J. Keast, J. Anstrom, A. L. Tyner, B. P. Brandhorst, and W. H. Klein. 1984. Novel proteins belonging to the troponin C superfamily are encoded by a set of mRNAs in sea urchin embryos. *Cell* **36**: 663.

Corces, V., A. Pellicer, R. Axel, and M. Meselson. 1981. Integration, transcription, and control of a Drosophila heat shock gene in mouse cells. *Proc. Natl. Acad. Sci. U.S.A.* **78**: 7038.

Davidson, E. H. 1976. *Gene Activity in Early Development.* Academic Press, New York.

Davidson, E. H., B. R. Hough-Evans, and R. J. Britten. 1982. Molecular biology of the sea urchin embryo. *Science* **217**: 17.

Davies, J. and D. I. Smith. 1978. Plasmid-determined resistance to antimicrobial agents. *Annu. Rev. Microbiol.* **32**: 469.

DiNocera, P. P. and I. B. Dawid. 1983. Transient expression of genes introduced into cultured cells of *Drosophila. Proc. Natl. Acad. Sci. U.S.A.* **80**: 7095.

Flytzanis, C. N., A. P. McMahon, B. R. Hough-Evans, K. S. Katula, R. J. Britten, and E. H. Davidson. 1985. Persistence and integration of cloned DNA in post-embryonic sea urchins. *Dev. Biol.* **108**: 431.

Goldberg, D. A., J. W. Posakony, and T. Maniatis. 1983. Correct developmental expression of a cloned alcohol dehydrogenase gene transduced into the Drosophila germ line. *Cell* **34**: 59.

Leahy, P. S., T. C. Tutschulte, R. J. Britten, and E. H. Davidson. 1978. A large-scale laboratory maintenance system for gravid purple sea urchins (*Strongylocentrotus purpuratus*). *J. Exp. Zool.* **204**: 369.

Lee, J. J., R. J. Shott, S. J. Rose, T. L. Thomas, R. J. Britten, and E. H. Davidson. 1984. Sea urchin actin gene subtypes. Gene number, linkage, and function. *J. Mol. Biol.* **172**: 149.

Lis, J., N. Costlow, J. deBangie, D. Knipple, D. O'Connor, and L. Sinclair. 1982. In *Heat shock: From bacteria to man* (ed. M. J. Schlesinger, M. Ashburner, and A. Tissières), p. 57. Cold Spring Harbor Laboratory, Cold Spring Harbor, New York.

McClay, D. R., G. W. Cannon, G. M. Wessel, R. D. Fink, and R. B. Marchase. 1983. In *Time, space, and pattern in embryonic development* (eds. W. R. Jeffery and R. A. Raff), p. 157. Alan R. Liss, New York.

McKnight, S. L. 1980. The nucleotide sequence and transcript map of the *Herpes simplex* virus thymidine kinase gene. *Nucl. Acids Res.* **8**: 5949.

McMahon, A. P., T. J. Novak, R. J. Britten, and E. H. Davidson. 1984. Inducible expression of a cloned heat shock fusion gene in sea urchin embryos. *Proc. Natl. Acad. Sci. U.S.A.* **81**: 7490.

McMahon, A. P., C. N. Flytzanis, B. R. Hough-Evans, K. S. Katula, R. J. Britten, and E. H. Davidson. 1985. Introduction of cloned DNA into sea urchin egg cytoplasm: Replication and persistence during embryogenesis. *Dev. Biol.* **108**: 420.

Scholnick, S. B., B. A. Morgan, and J. Hirsh. 1983. The cloned dopa decarboxylase gene is developmentally regulated when reintegrated into the *Drosophila* genome. *Cell* **34**: 37.

Spradling, A. C. and G. M. Rubin. 1982. Transposition of cloned P elements into *Drosophila* germ line chromosomes. *Science* **218**: 341.

———. 1983. The effect of chromosomal position on the expression of the *Drosophila* xanthine dehydrogenase gene. *Cell* **34**: 47.

Voellmy, R. and D. Rungger. 1982. Transcription of a *Drosophila* heat shock gene is heat induced in *Xenopus* oocytes. *Proc. Natl. Acad. Sci. U.S.A.* **79**: 1776.

Fate of DNA Injected into Xenopus Eggs and in Egg Extracts: Assembly into Nuclei

JOHN NEWPORT AND DOUGLASS J. FORBES
Department of Biology
University of California, San Diego
La Jolla, California 92093

OVERVIEW

We report here that naked DNA from bacteriophage lambda is assembled into nuclear structures when this DNA is either injected into *Xenopus* eggs or mixed with extracts made from these eggs. These induced nuclei are structurally identical to "real" nuclei in that they are surrounded by a double membrane envelope that contains nuclear pores and they contain the major nuclear lamin protein. These reconstituted nuclei are functionally similar to normal nuclei in that they transport molecules normally found in the nucleus, breakdown during mitosis, and replicate their DNA. These results suggest that DNA injected into eggs from other animals might also be assembled into nuclear structures.

INTRODUCTION

During the first 7 hours following fertilization, the *Xenopus* egg undergoes 12 synchronous cell divisions without cell growth or measurable RNA transcription (Hara 1977; Newport and Kirschner 1982a). This rapid cleavage period terminates abruptly following the twelfth cleavage division (midblastula transition), at which time the cell cycle slows down, transcription is activated, and cells become motile (Bachvarova and Davidson 1966; Newport and Kirschner 1982b). The rapid cell cycle following fertilization is made possible because the egg contains large stores of structural and enzymatic components necessary for division. The egg contains a pool of histones sufficient for assembling 20,000 nuclear equivalents of DNA into chromatin (Laskey et al. 1977). Furthermore, when DNA is injected into the egg, it is rapidly assembled into native chromatin, demonstrating that the components necessary for this assembly process are present within the egg. It has been recently shown that, when demembranated sperm nuclei are added to a cytoplasmic extract made from activated eggs, the nuclei reacquire a nuclear envelope, indicating the presence of a pool of envelope components in the egg (Lohka and Masui 1983). A number of other components including small nuclear RNAs (Forbes et al. 1983a), small nuclear RNA proteins, DNA polymerase, tubulin, and ribosomes have also been shown to be present in large amounts. In this paper, we show that most of the structural components of the nucleus are also stored in excess within the egg. Furthermore, we show that these components assemble into normal nuclei when presented with a DNA template (Forbes et al. 1983b). This assembly process is not DNA sequence-specific and occurs in an in vitro cytoplasmic extract obtained from *Xenopus* eggs. This system should be useful for studying important aspects of nuclear assembly and perhaps for packaging DNA into a stable entity for transformation studies.

RESULTS

Bacteriophage λ DNA Acts as a Template for Nuclear Formation

When linear bacteriophage λ DNA is injected into unfertilized enucleated *Xenopus* eggs, it undergoes a series of physical transformations (Fig. 1). Initially the DNA appears as an unorganized filamentous network. However, within 30–45 minutes this unorganized network condenses into shorter thicker filaments. Following this initial condensation, the DNA becomes further condensed into densely packed spheres (45–70 minutes postinjection). By 90 minutes after injection into the egg, the densely packed spheres appear to have decondensed and the DNA is now encapsulated within a membranous envelope. These DNA-containing vesicles are

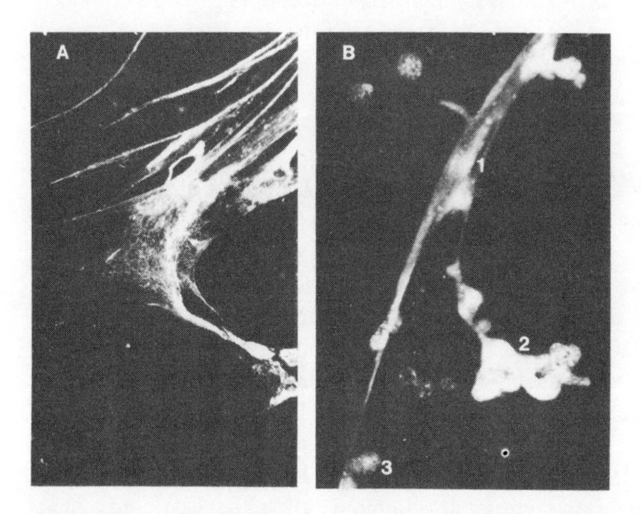

Figure 1
Fate of bacteriophage λ DNA injected into *Xenopus* eggs: Bacteriophage λ DNA (5–8 ng) was injected into unfertilized *Xenopus* eggs. The injected eggs were lysed at different times in the DNA-specific fluorescent dye bisbenzimide and 3.7% formaldehyde and then observed via fluorescence microscopy. (*A*) DNA in an egg lysed immediately after injection showing the DNA as a disperse fibrous array; (*B*) DNA in an egg lysed 90 min after injection. The three stages of DNA rearrangement are shown: (*1*) partially condensed DNA; (*2*) highly condensed and partially spherical DNA; and (*3*) membrane-enclosed nuclei. Reprinted, with permission, from Forbes et al. (1983a).

heterogeneous in size ranging from 1–20 micrometers in diameter. However, the final size of a given vesicle appears to depend directly on how much DNA is contained within it. Optical sectioning through these spheres demonstrated that the bulk of the DNA was attached to the membrane, i.e., the centers of the spheres did not contain appreciable amounts of DNA. Thus it appears from these results that bacteriophage λ DNA injected into *Xenopus* eggs acts as a template for the formation of nucleus-like structures.

In order to demonstrate that the nuclei seen via fluorescence are derived from injected λ DNA and not from division of the endogenous maternal nucleus, eggs were manually enucleated with a needle and injected with λ DNA. Enucleated eggs injected with DNA formed nuclei whereas those injected with buffer alone did not. Furthermore, when ^{3}H-labeled λ DNA was injected into eggs, autoradiographic analysis of the nuclei formed showed that it was the λ DNA that was contained in the nuclei. These results show that the nuclei formed were the result of injection of λ DNA.

Nuclei Formed Around λ DNA Contain Nuclear Lamins, Membranes, Nuclear Pores, and Small Nuclear RNPs

The envelope of a normal somatic cell nucleus is made up of a double membrane, interrupted by a complex of proteins making up the nuclear pores (Fig. 2). The inner nuclear membrane is lined by a proteinaceous layer composed of one or more of the lamin proteins A, B, and C (Gerace and Blobel 1980). If the structures formed around λ DNA injected into *Xenopus* eggs are "true" nuclei, then they should contain a double membrane at their periphery, nuclear pores, and a nuclear lamina.

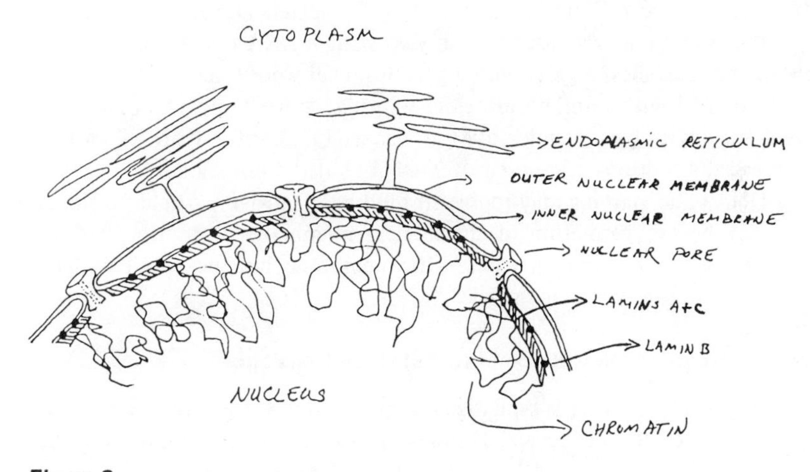

Figure 2
Schematic diagram of the organization of a typical nucleus

Nuclear structures which formed around λ DNA injected into *Xenopus* eggs were examined via electron microscopy. The λ DNA-induced nuclei were found to possess a double membrane identical to that found around normal *Xenopus* embryonic nuclei. Large cytoplasmic structures such as mitochondria and yolk granules were excluded from the inside of the nuclei by this envelope. Furthermore, these nuclear membranes were periodically interrupted by typical nuclear pores. The nuclear envelope and nuclear pores were indistinguishable from those present around normal embryonic nuclei, both in appearance and in the spacing of pores within the envelope. Thus, λ DNA-induced nuclei appear to be "real" nuclei in terms of envelope and pore content.

In order to determine if these nuclei contained the nuclear lamin proteins, we used a highly specific antibody against the lamins to stain the nuclei (McKeon et al. 1983). Following staining with the anti-lamin antibody, the nuclei were stained with a rhodamine-labeled secondary antibody. Lambda DNA-containing nuclei treated in this manner were observed to have a brightly staining fluorescent outer perimeter enclosing a central core of DNA. The staining pattern was identical to the pattern present in embryonic nuclei. This result demonstrates that the nuclei that form around λ DNA injected into eggs contain the lamin proteins present in normal embryonic nuclei. Thus these induced nuclei appear normal in terms of envelope structure, pore content, and presence of nuclear lamins.

Although the "λ nuclei" resembled normal nuclei in many structural characteristics, we wished to know whether they contained nuclear molecules that were not an inherent part of the nuclear structure itself. The small nuclear RNA molecule, U1, is contained in a ribonucleoprotein particle (snBNP) and this particle, although normally nuclear, is thought to be involved in an enzymatic role in the nucleus rather than a structural role. It has been shown in amoeba that small nuclear RNAs can shuttle between two nuclei of a heterokaryon (Goldstein and Ko 1974). To determine whether the nuclei reconstituted in vivo with λ DNA could bind or transport small snRNP particles, eggs containing such nuclei were fixed, sectioned in a cryostat, and stained with immunofluorescent anti-U1 snRNP antisera. (The DNA was counterstained with bisbenzimide dye). It was seen that anti-U1 snRNP antisera did stain spherical structures composed of λ DNA, but did not stain less organized masses of λ DNA. The staining may represent binding of snRNP particles to the organized λ DNA before formation of the nuclear membrane or transport of the particles into the nucleus after formation of the membrane, but the snRNP antigens are clearly present in the synthetic nuclei.

Nuclei Formed Around λ DNA Respond to Cell Cycle Regulation

The early embryonic cell cycle is biphasic containing only S (DNA replication) and M (mitosis) phases of the cell cycle. During mitosis nuclei disassemble their envelopes and the DNA condenses into metaphase chromosomes. Masui and coworkers have demonstrated that unfertilized, nonactivated *Xenopus* eggs contain a cytoplasmic factor called cytostatic factor (CSF) which, when injected into fertilized eggs, causes these cells to arrest in mitosis (Meyerhof and Masui 1979). In order to determine if the nuclei formed around λ DNA respond to the molecular signals regulating nuclear breakdown and chromosome condensation during mitosis, the following was done: Lambda DNA was injected into fertilized *Xenopus* eggs and incubated for 90 minutes. During this period the DNA was packaged into nuclei. Following this, the eggs were arrested in mitosis by injecting cytoplasm from unfertilized eggs containing CSF activity. When we examined these mitotically arrested eggs for λ DNA-containing nuclei, we found that they were no longer present. Instead, the λ DNA was now found in a highly condensed configuration similar to metaphase chromosomes and lacking any evidence of a nuclear envelope. This experiment demonstrates that the nuclei containing λ DNA respond to the molecular signals that induce endogenous nuclei to break down their nuclear envelopes and condense the chromosomes during mitosis.

Nuclei Form Around λ DNA in Cell-free Extract Made from Xenopus Eggs

In order to investigate the process of nuclear assembly at the biochemical level, it was determined that an in vitro nuclear regrowth system would be highly advantageous. To obtain such a system, we made extracts from *Xenopus* eggs that reconstitute nuclear structures around added λ DNA. Such extracts are made from activated *Xenopus* eggs which have been lysed and centrifuged to remove both pigment and yolk granules. When λ DNA is added to such a cleared egg extract in the presence of an ATP-regenerating system, the DNA becomes packaged into nuclear structures. These in vitro reconstituted nuclei form in a highly synchronous process similar to formation of nuclei around injected DNA in vivo. The added DNA undergoes an initial condensation resulting from its assembly into chromatin. Following this the DNA becomes further compacted into dense spherical structures. Based on indirect immunofluorescent staining with anti-lamin antibody and electron microscopic observations, these dense spherical structures lack both nuclear lamins and nuclear envelope. A likely possibility is that these dense intermediates are held in their highly condensed conformation by a scaffold-like protein similar to that described by Laemmli and coworkers. The condensed intermediates then decondense in a process during which lamin proteins and nuclear envelope become associated with them. The end product of nuclear assembly in vitro is λ DNA packaged within a nucleus bounded by a double membrane containing nuclear pores, and lined by a nuclear lamina. Typically, 1000 eggs yield 0.5 ml of extract capable of packaging approximately 5.0 μg of λ DNA.

Functionally, these extracts are very active in DNA synthesis. In Figure 3, the incorporation of radioactive dCTP into λ DNA after addition to such an extract is graphed. Based on the specific activity of the DNA at 70 minutes, it would appear that at least 50% of the DNA has been replicated. Such replication does not appear

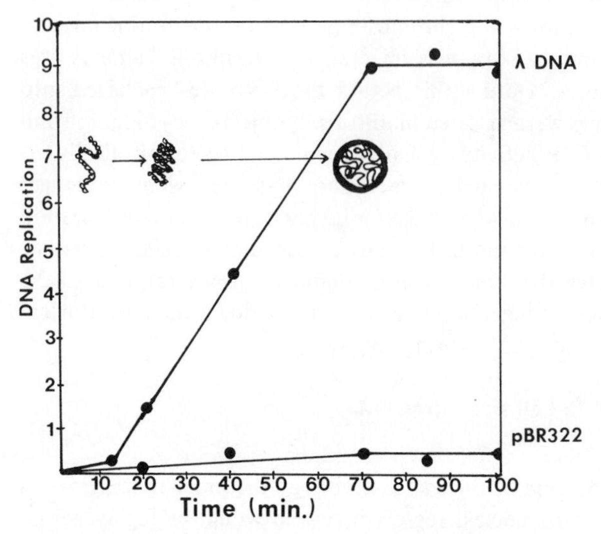

Figure 3
Lambda DNA replication in extracts made from *Xenopus* eggs. 500 ng of λ DNA were mixed with 50 μl of egg extract and 50 μCi of labeled dCTP. Aliquots were removed at various times; DNA purified, run on an agarose gel, and the DNA bands were cut out and counted. At the plateau (70 min) the specific activity of the λ DNA was 10^6 cpm/μg. The figure also shows the replication of pBR322 DNA (which replicates with 1/20th the efficiency of linear λ DNA) and the conformation of the λ DNA during replication as determined visually with fluorescence microscopy. The majority of replication (80%) occurs during a period when the DNA is binding nuclear lamins and membrane.

to be due to repair as it is sensitive to the DNA polymerase α-specific inhibitor, aphidicolin. Whether this replication is due to a single round of synthesis or repeated rounds of synthesis on a small fraction of the total DNA is currently under investigation.

Transport of Nuclear Molecules into Synthetic Nuclei Formed In Vitro

In order to test whether the nuclei formed in vitro were capable of transporting nuclear molecules, radioactive small nuclear RNAs and tRNAs were added to synthetic nuclei formed in vitro. Specifically, λ DNA was added to an extract of *Xenopus* eggs, and sufficient time allowed for nuclei to form (as determined visually). Radioactive *Xenopus* embryonic RNA, containing U1, U2, U4, U5, and tRNA was added to the mixture. After 50 minutes, the nuclei were pelleted and the RNA extracted from both the nuclear and supernatant fractions and separated on an acrylyamide gel. (The snRNP proteins are present and available for binding to the snRNAs in the extract).

The nuclear fraction was enriched in the snRNAs U1, U2, U4, and U5 whereas the supernatant fraction contained an excess of tRNAs. These results are consistent with the synthetic nuclei either transporting the snRNPs or binding them in some abortive transport event. Further experiments are needed to rule out other possibilities, such as preferential degradation of one class of RNAs in alternate fractions. The reconstitution system should prove quite useful in defining the requirements for nuclear transport.

Transcription in Reconstituted Nuclei

When naked DNA is injected into the cytoplasm of a *Xenopus* oocyte, it is rapidly degraded; whereas if it is injected into the nucleus of the oocyte (germinal vesicle), it is stable (Wyllie et al. 1978). However, when λ DNA-containing nuclei, assembled in vitro, are injected into the cytoplasm of an oocyte, the DNA packaged within the nuclei is stable. We are currently investigating the transcriptional properties of these in vitro nuclei when they are injected into oocytes. It is possible that by pre-assembling the DNA into a nucleus, transcriptional efficiency of the DNA may increase substantially over the 1-3% efficiency normally observed when DNA is injected into the oocyte germinal vesicle.

SUMMARY

We have demonstrated that when DNA is injected into *Xenopus* eggs, it becomes enclosed in a structure that appears to be identical to a normal embryonic nucleus in that it contains a double membrane nuclear envelope, nuclear pores, and lamin content. Functionally, these reconstituted nuclei are normal in that they break down their nuclear envelopes and condense their DNA during mitosis. Furthermore, they appear active in the selective transport of the normal nuclei-localized, small nuclear RNA particles. We have also characterized an in vitro system for nuclear reconstitution that assembles nuclei around naked DNA. These in vitro nuclei replicate their DNA efficiently, transport proteins, and are stable in the cytoplasm of *Xenopus* oocytes.

These results demonstrate that the *Xenopus* egg contains a large excess of stored, unassembled, nuclear, structural components which will assemble "spontaneously" around a DNA template. Furthermore, such assembly around bacteriophage λ DNA strongly suggests that nuclear assembly is not dependent on specific DNA sequence information. Because a number of animals including fruit flies, sea urchins, and, to some extent, mice also undergo a period of rapid cleavage following fertilization, it is likely that they also contain pools of unassembled nuclear components that will assemble around injected DNA.

REFERENCES

Bachvarova, R. and E. Davidson. 1966. Nuclear activation at the onset of amphibian gastrulation. *J. Exp. Zool.* **163**: 285.

Forbes, D., M. Kirschner, and J. Newport. 1983a. Spontaneous formation of nucleus-like structures around bacteriophage DNA microinjected into *Xenopus* eggs. *Cell* **34**: 13.

Forbes, D., T. Kornberg, and M. Kirschner. 1983b. Small nuclear RNA transcription and assembly in early *Xenopus* development. *J. Cell Biol.* **97**: 62.

Gerace, L., and G. Blobel. 1980. The nuclear envelope lamina is reversibly depolymerized during mitosis. *Cell* **19**: 277.

Goldstein, L., and C. Ko. 1974. Electrophoretic characterization of shuttling and nonshuttling small nuclear RNAs. *Cell* **2**: 259.

Hara, K. 1977. The cleavage pattern of the axolotl egg studied by cinematography and cell counting. *Wilhelm Roux' Arch. Entwicklungsmech Org.* **181**: 73.

Laskey, R., A. Mills, and N. Morris. 1977. Assembly of SV40 chromatin in a cell-free system from *Xenopus* eggs. *Cell* **10**: 237.

Lohka, M., and Y. Masui. 1983. Formation in vitro of sperm pronuclei and mitotic chromosomes induced by amphibian ooplasmic components. *Science* **220**: 719.

McKeon, F., D. Tuffanelli, K. Fukuyama, and M. Kirschner. 1983. An autoimmune response directed against conserved determinants of nuclear envelope proteins in a patient with linear scleroderma. *Proc. Nat. Acad. Sci. U.S.A.* **80**: 4374.

Meyerhof, P., and Y. Masui. 1979. Properties of a cytostatic factor from *Xenopus* eggs. *Dev. Biol.* **72**: 182.

Newport, J. and M. Kirschner. 1982a. A major developmental transition in early *Xenopus* embryos: II. Control of the onset of transcription. *Cell* **30**: 687.

———. 1982b. A major developmental transition in early *Xenopus* embryos: I. Characterization and timing of cellular changes at the midblastula stage. *Cell* **30**: 675.

Wyllie, A., R. Laskey, J. Finch, and J. Gurdon. 1978. Selective DNA conservation and chromatin assembly after injection of SV40 DNA into *Xenopus* oocytes. *Dev. Biol.* **64**: 178.

Extrachromosomal DNA Transformation of C. elegans

DAN T. STINCHCOMB,* JOCELYN SHAW,† STEPHEN CARR,† AND DAVID HIRSH†
*Department of Cellular and Developmental Biology
Harvard University
Cambridge, Massachusetts 02138
†Department of Molecular, Cellular, and Developmental Biology
University of Colorado
Boulder, Colorado 80309

OVERVIEW

We have recently demonstrated that DNA can be introduced into the germline of the worm by microinjection; 10% of the injected worms give rise to progeny that maintain the foreign DNA. The exogenous DNA is present as a high molecular weight array. The array appears to be created by ligation of linear molecules and/or by recombination between supercoils. This high molecular weight foreign DNA is heritable: on average, 50% of the progeny of a transformed self-fertilizing hermaphrodite will still carry the exogenous sequences. Cytological evidence indicates that the array of foreign sequences is extrachromosomal. In situ hybridization experiments demonstrate that the extrachromosomal sequences can be lost during the growth of the organism giving rise to mosaic animals. Approximately one-half of the transformed animals are mosaic; the remainder carry foreign DNA in all of their cells. Loss of the foreign DNA may be due to errors in replication or segregation of the exogenous sequences. No worm DNA sequences are required in the injected plasmid for the formation of these extrachromosomal arrays. To assess expression, we constructed a gene fusion between the 5′ and 3′ sequences of a *C. elegans* collagen gene and the bacterial coding sequences for β-glucuronidase (*uid*A). After injection, the gene fusion is expressed as part of a high molecular weight array. Thus, we may contemplate the use of this unusual mode of transformation to study developmental regulation in *C. elegans.*

INTRODUCTION

The small nematode *Caenorhabditis elegans*, has been widely used for studies of developmental genetics. Not only is the organism easy to culture and manipulate genetically (Brenner 1974), but the entire cell lineage from the fertilized egg to the adult has been detailed (Sulston and Horvitz 1977; Kimble and Hirsh 1979; Sulston et al. 1983). This virtually invariant cell lineage and the relatively simple anatomy of *C. elegans* allows detailed comparison of mutant and wild-type developmental patterns. Many mutants have been isolated and characterized that are altered in their cell lineages. For example, mutations in *par-1*, a gene that is maternally required for proper embryogenesis, change the pattern of cell divisions in early development (K. Kemphues, J. Priess, and D. Hirsh, in prep.). Analysis of mutations in the *lin-12* gene suggests that the level of gene activity controls how a particular cell chooses between two cell fates (Greenwald et al. 1983). Another class of lineage mutants affects the timing of cell division patterns during development (Ambros and Horvitz 1984). Other genes have been characterized that are central to the unfolding of developmental pathways, to the expression of specific behaviors, or to the differentiation of particular tissues and cells. Our long-term goal is to meld molecular techniques to the detailed physiological and genetic characterization of genes that control important developmental processes.

Caenorhabditis elegans is also amenable to analysis in molecular detail. Large populations of worms can be grown synchronously, permitting the purification of stage-specific macromolecules. Due to its simple cellular anatomy and transparency, macromolecules can be precisely localized with antibody or nucleic acid probes. Several genes encoding abundant proteins or RNAs have been isolated using recombinant DNA technology (MacLeod et al. 1981; Kramer et al. 1982; Files et al. 1983). Recently, progress has been made towards cloning particular loci by chromosome walking from genetically mapped restriction fragment polymorphisms (I. Greenwald et al., pers. comm.). DNA transformation is the missing linchpin in the molecular biology of *C. elegans.* For the genes already isolated, transformation is required to analyze the sequences necessary for proper gene regulation during development. Mutations can be made in regulatory sequences in vitro; the alterations can be tested for function by reintroducing the sequences into the worm. In addition, DNA transformation may permit the isolation of new loci. For instance, transformation with a *C. elegans* hybrid molecule may complement a mutant of interest. The DNA encoding the complementing activity might be reisolated from the transformed genome by virtue of its bacterial moiety. Thus, a DNA transformation system would permit identification of DNA encoding almost any locus in the worm genome, including those known to be crucial to developmental processes. Towards this end, we have been developing the techniques and investigating the mechanism of DNA transformation of *C. elegans.*

RESULTS

Microinjection

The first requirement of a DNA transformation system is a means of introducing DNA into the germline of the organism. We microinjected DNA into the distal arm of the adult hermaphrodite gonad. In the distal arm, the ovary is a syncytium: circumferential nuclei share a central core of cytoplasm. Figure 1 shows the injection of several picoliters of a solution of supercoiled DNA into the distal gonad arm of

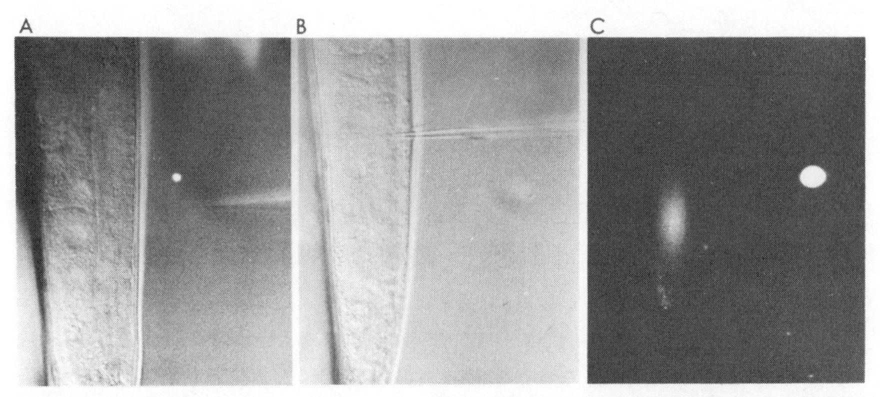

Figure 1
Microinjection of DNA. Worms were anesthetized in 0.1% tricaine, 0.01% tetramisole, and were lined up against a cover slip under an oxygen-permeable halocarbon oil. *Panels A and B* show the posterior of an adult hermaphrodite before and after penetration of the cuticle with the microinjection needle. The tip of the needle (0.1 μ to about 2.0 μ) was backfilled by capillary action with 0.5 mg/ml supercoiled DNA in 1.0 mM KCl, 0.1 mM $K_{1.5}PO_4$, pH 7 and 1 mg/ml fluorescein-conjugated dextran. The remainder of the needle is filled with oil and attached to a screw-type Hamilton syringe. By applying pressure with the syringe, 1–20 pl of the solution was injected. Successful injection was monitored by fluorescence of the coinjected dextran as shown in *Panel C.*

C. elegans. The techniques involved only minor modification of Kimble's RNA injection procedures (Kimble et al. 1982). The worms are anesthetized, lined up along the edge of a cover slip, and covered with an oxygen-permeable halocarbon oil. Hermaphrodites remained viable and immobile for several hours under these conditions. *Panel A* (Fig. 1) shows the posterior of an anesthetized hermaphrodite and *panel B* shows the penetration of the distal gonad arm with a freshly broken glass needle. *Panel C* is an epifluorescent image of a successful injection. The diffuse fluorescent material is coinjected fluorescein-conjugated dextran present in the distal gonad arm; the bright spot is a drop of the same material present on the microinjection needle. In this fashion, several worms can be injected with DNA in the course of a few hours. Greater than 80% of the worms survive and produce progeny from the injected gonad.

DNA Transformation

To determine if the exogenously added DNA was heritable, we examined the progeny of the injected worms for foreign DNA sequences by hybridization techniques. Groups of progeny were propagated for several generations, harvested, and had their DNA purified. Figure 2 shows the hybridization of ^{32}P-labeled pBR322 sequences to DNA isolated from the progeny of four injected worms. The first batch of progeny from injection 3 shows evidence of exogenously added DNA sequences.

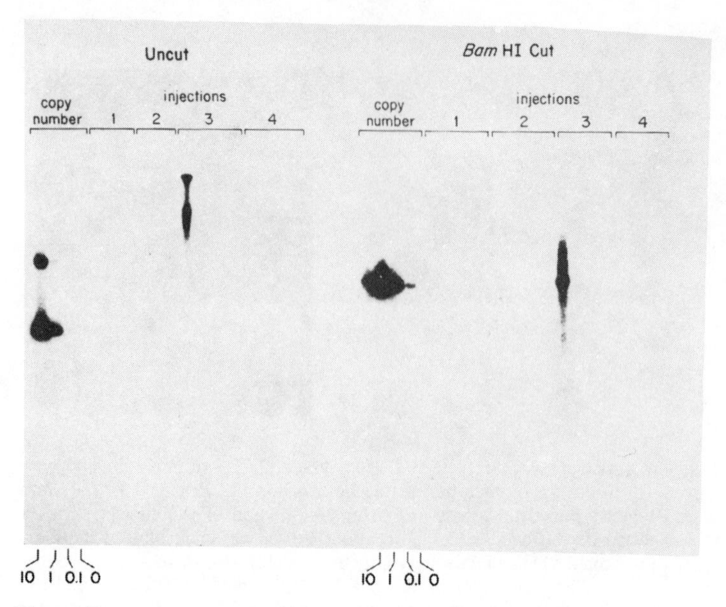

Figure 2
DNA transformation. Four worms were successfully injected with a supercoiled plasmid bearing a 3.4-kbp *C. elegans* DNA fragment. The progeny of each worm were collected in groups of about 40 worms and were propagated for 2–3 generations. Each batch of worms was harvested; their DNA was purified, sized by agarose gel electrophoresis, and transferred to the nitrocellulose paper. The filters were hybridized with ^{32}P-labeled plasmid and washed; autoradiography was then performed. In *Panel A*, the worm DNA was electrophoresed without prior digestion. On the right, the DNA was cleaved with the restriction endonuclease *Bam*HI. *Bam*HI cuts once in the plasmid sequences. The copy number controls shown are reconstructions, adding small amounts of plasmid to wild-type worm DNA to emulate 10, 1, 0.1, and 0 copies per haploid genome.

What is the structure of the foreign DNA? When the DNA isolated from the transformed worms was subjected to electrophoresis without prior endonuclease digestion, the foreign DNA was found in the high molecular weight fraction (Fig. 2, gel labeled *Uncut*). In the gel on the right in Figure 2, the transformant's DNA was digested with a restriction enzyme that cleaves once in the original injected plasmid. The exogenous sequences now migrated heterogeneously; however, a predominant band comigrated with the linear injected plasmid (Fig. 2; compare injection 3 to *copy number* controls). These data suggest that most of the foreign DNA is present as a large head-to-tail tandem array of plasmid sequences. High molecular weight concatameric foreign DNA was detectable in ten out of 114 injections. These tandem arrays could arise from recombination between injected circular molecules. Alternatively, an amplification mechanism might be involved.

Injection of Linear DNA

When transformed into yeast, linear molecules will homologously recombine with the yeast chromosomes (Orr-Weaver et al. 1981). Linear DNA can also be used to transform *C. elegans.* A plasmid carrying a fragment of the *unc-54* gene was cut at a unique site in the worm sequence. After injection, a transformed strain was obtained as described above. If the linear molecule had recombined homologously, it would have disrupted the *unc-54* locus resulting in a distinctive uncoordinated phenotype; none was observed amongst the transformants or their progeny. The structure of the foreign DNA carried in the transformants was different from the structures observed after injection of supercoiled molecules. When subjected to electrophoresis without digestion, the hybridizing sequences were of high molecular weight. However, when the foreign DNA was digested with a restriction site that cuts once in the injected molecule, three predominant bands were observed. The bands corresponded to those expected for head-to-tail, head-to-head, and tail-to-tail ligation of the injected linear molecule (data not shown). The single strand cohesive ends of the injected molecules seem vulnerable to digestion in *C. elegans*; a cohesive end restriction site is not reconstructed after ligation (data not shown). Transformants produced by injection of linear molecules or supercoils behaved identically in all the experiments described below.

Stability of High Molecular Weight Transformants

Transformation by the formation of such high molecular weight concatamers does not require *C. elegans* DNA sequences. Four of the plasmids we used to transform contained no extensive homology with the worm genome. We tested the stability of one of these nonhomologous transformants (injected with the plasmid pSV2neo) (Southern and Berg 1982) over several generations as follows: Single self-fertilizing hermaphrodites were propagated for three generations and then one was identified as a transformant by hybridization analysis (e.g., worm isolate 1 in Table 1). A large number of the transformant's third generation progeny were individually transferred to separate plates. The percentage of these progeny which still carry exogenous DNA was extrapolated to the stability per generation in the last column of Table 1. The exogenous high molecular weight sequences were lost at a surprisingly high rate; on average, 50% of the progeny of a transformed worm still carried the high molecular weight array. Occasionally, we found transformants with higher stabilities of 80-90% per generation (see clone 1.1.3.3 in Table 1); however the high stability was a transient phenomenon. We have maintained strains carrying the foreign DNA for dozens of generations and we have yet to find a transformant that is completely stable.

We examined several progeny that still carried the exogenous sequences in more detail. DNA prepared from these worms was digested with restriction endonucleases

Table 1
Stability of Foreign DNA

Isolated worm	Number of generations	Fraction positive	Stability/ generation (%)
1	3	1/24	35
1.1	3	7/48	53
1.1	6	1/48	52
1.1.1	3	6/50	49
1.1.2	3	9/48	57
1.1.3	3	5/25	59
1.1.3.2	3	4/25	54
1.1.3.3	3	18/25	90

and compared to DNA from the parental transformant. All showed indistinguishable patterns of hybridization, both in quantity of hybridization and in the spectra of minor bands displayed (Fig. 3). Progeny that lacked hybridization contained fewer than 0.1 copies of exogenous sequences per haploid genome (data not shown). Thus, the foreign DNA segregates as a single unit and is stably maintained over many generations.

Copy Number of the Foreign DNA

To determine the number of copies of the foreign sequences present per transformed worm, we first isolated DNA from a population of transformants. We then determined the percentage of transformed worms present in the population. By quantitating the hybridization of plasmid sequences to the DNA and correcting for the number of transformed worms in the population, we could calculate the plasmid copy number per transformed worm. Purified worm DNA was spotted onto triplicate nitrocellulose filters and hybridized with excess ^{32}P-labeled bacterial plasmid sequence, a single-copy worm sequence, and a fourfold repeated worm sequence. Hybridization was quantitated by scintillation counting and standardized for the hybridization efficiencies of the different probes. The average copy number obtained for the fourfold repeated actin genes was only 3.0 suggesting a routine underestimate. The transformants show high copy numbers of 80–300 foreign sequences per haploid genome (Table 2).

Genetic Analysis of the Transformants

The low stability of the transforming DNA suggests non-Mendelian segregation. Alternatively, the foreign DNA could impart a deleterious phenotype to the worms. In the stability experiments described above, third generation progeny were care-

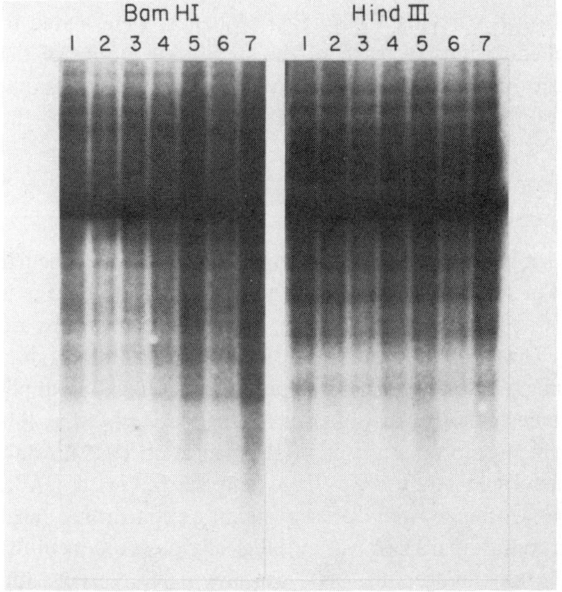

Figure 3
Equivalent DNA in progeny. During a standard stability test, six progeny of a single transformed worm were identified that still carried the exogenous DNA sequences. DNA preparations from these third generation progeny were cleaved with two different restriction endonucleases that cut once in the original injected plasmid (*Bam*HI and *Hin*dIII, as shown). The digested DNA was size-fractionated in an agarose gel, transferred to a nitrocellulose filter, and hybridized with ^{32}P-labeled plasmid sequences. An overexposure of the subsequent autoradiograph is shown here. Lane *1* contains DNA isolated from the original parent; lanes *2*-*7* are the six progeny. No major differences in the content of foreign DNA sequences could be detected.

Table 2
Copy Number

Population	Copy number		Fraction positive	Copy number/ transformant
	Actin	pBR		
pSV2neo-1	2.8	8.4	1/24	200
-1.1.1	4.5	36.3	6/50	300
-1.1.2	3.1	32.7	9/48	170
-1.1.3	2.4	32.8	5/25	160
-1.1.3.2	3.1	37.5	4/25	230
-1.1.3.3	2.1	57.7	18/25	80

fully screened for any detectable reduction in brood size, growth rate, or fertility; none was observed. Transformants carrying the foreign sequences were mated to strains bearing morphological mutations on each of the six linkage groups of the worm. Heterozygous worms carrying the foreign DNA were identified. Progeny that were homozygous for each of the morphological mutations were scored for the presence of the foreign DNA. No linkage was detected.

Extrachromosomal Foreign DNA

To further explore the cellular location and the instability of the high molecular weight arrays, we analyzed the foreign DNA cytologically. The mature oocyte is arrested in diplotene of the first meiosis. At this stage, the six linkage groups are paired and highly condensed. They can be visualized as six brightly fluorescing dots by staining with diamidinophenylindole (DAPI). Independently segregating duplications of a worm chromosome (free duplications) can also be visualized at this stage as diffusely staining extrachromosomal material (Herman et al. 1979). Adult hermaphrodites from a transformed strain were fixed and stained with DAPI. Oocytes in several of the adult worms showed diffusely staining extrachromosomal material (Fig. 4). The extrachromosomal DNA was cytologically observed in adults if and only if their progeny bore exogenous DNA as shown by hybridization analysis (see legend to Fig. 4).

Mosaic Transformants

To localize the extrachromosomal arrays in the developing transformants, we hybridized ^{35}S-labeled sequences directly to worm tissues. Embryos and bissected adults were fixed and prepared for hybridization using methods developed by D. Albertson (pers. comm.). Only those tissues that were well separated from the cuticle or the egg shell were considered to avoid problems of probe accessibility. As a control for probe sensitivity and accessibility, we hybridized tissues with ^{35}S-labeled ribosomal DNA. The rDNA probe reproducibly hybridized to all of the nuclei in the gonadal and intestinal tissue of the adult and to all embryonic nuclei. ^{35}S-labeled pBR322 did not hybridize to uninjected worms. However, the pBR322 sequences hybridized dramatically to the transformed worm tissues. Half of the embryos and many of the adults displayed hybridization to all nuclei. Some adults showed hybridization to somatic intestinal nuclei but no hybridization to the gonad. Other adults showed mosaic patterns of hybridization to somatic and/or germline nuclei. Likewise approximately 50% of the embryos were mosaic (Fig. 5). The full spectrum of mosaicism was seen: from hundred cell embryos in which two nuclei hybridized, to embryos in which all but two nuclei hybridized. Thus, the instability of the extrachromosomal array applies to both somatic and germline lineages.

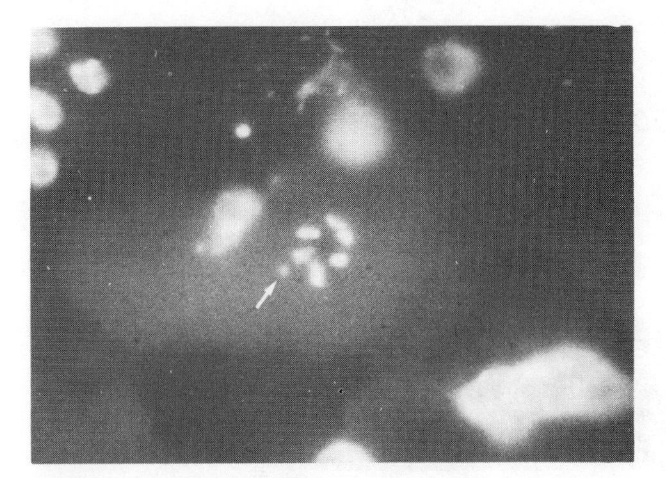

Figure 4
Extrachromosal arrays. Single adult hermaphrodites were picked from a transformed strain. The worms were first transferred to agar plates where they laid 20–50 eggs. These same adults were then individually transferred to slides, bissected to display the gonads, and were frozen, fixed, and stained with the fluorogenic DNA binding agent, DAPI. The slides were examined with epifluorescent illumination. The figure shows a single oocyte surrounded by several gonadal sheath cells. The oocyte's chromosomes are seen as six brightly staining pairs of dots. The arrow indicates extrachromosomal material. In the meantime, the progeny of these adults were allowed to propagate; DNA was prepared from the populations and screened for the presence of foreign DNA. As mentioned in the text, the presence of extrachromosomal material as shown here correlated with the presence of foreign DNA as determined by the dot blots. In one adult that must have contained the foreign DNA sequences, we failed to observe any autonomous DNA; the instability of the high molecular weight arrays and the low number of well-prepared oocytes per worm may have contributed to this failure.

Expression of Foreign DNA

We are constructing fusions between bacterial coding sequences and isolated worm genes in order to explore the expression of the exogenously added DNA. One such construction fused the bacterial *uidA* structural gene to 5′ and 3′ fragments of the *C. elegans col-1* gene. *uidA* encodes the enzyme β-glucuronidase. Histochemical techniques had been developed to stain worms for the presence of the β-glucuronidase activity and mutants of *C. elegans* were isolated that were deficient for the enzyme (P. Bazzicalupe, R. Jefferson and D. Hirsh, unpubl.). The *col-1* gene is a member of the collagen multigene family (Kramer et al. 1982). Its expression is tissue-specific and temporally characterized (Kramer et al. 1985). A 570 bp 5′ fragment of *col-1* containing the transcriptional start site and initiating AUG codon was ligated in frame to the complete *uidA* coding sequence. A 350 bp 3′ region of *col-1* containing a splice and poly-A addition site was inserted downstream. The

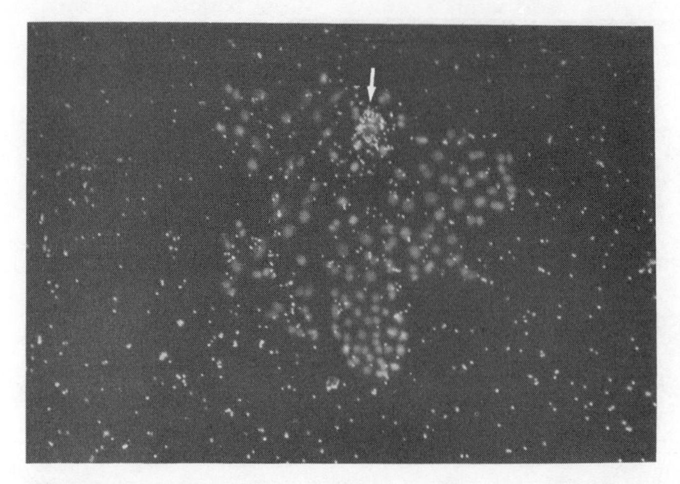

Figure 5
Mosaic transformants. The methods used for in situ hybridization were simple modifications of procedures developed by D. Albertson (pers. comm.). After fixation and digestion with RNAse, the tissues were hybridized with ^{35}S-labeled plasmid DNA, prepared by nick-translation. Hybridization and subsequent washing were carried out in the usual formamide conditions. The slides were coated with emulsion, allowed to expose, developed, and stained with DAPI. The figure shows an embryo with approximately one hundred nuclei (grey, diffuse spots) as viewed with darkfield and epiflourescent illumination. Under darkfield illumination, the exposed grains appear as bright dots. A cluster of hybridization is seen centered over two of the one hundred nuclei in the upper portion of the embryo.

chimeric gene was injected as supercoiled DNA into the β-glucuronidase null mutant. A transformant containing the foreign DNA sequences was obtained as described above. Progeny of the transformant were tested for presence of both the exogenous sequences and the β-glucuronidase activity. All progeny that contained the DNA also contained the enzymatic activity (data not shown). The activity has cosegregated with the DNA for many generations. Experiments are in progress to define the tissue and temporal expression of the gene fusion.

DISCUSSION

We have transformed the genome of *C. elegans* by microinjecting DNA into the cytoplasm of the syncytial gonad. Injection of DNA into the cytoplasm of mammalian cells leads to rapid degradation (Capecchi 1980). In *C. elegans*, cytoplasmic degradation does not appear to be a limiting factor for transformation; transformants have been obtained in wild-type and nuclease-deficient gonads at comparable frequencies. Foreign DNA was detected amongst the progeny in approximately 10% of the injections. We have not investigated the factors affecting the frequency of transformation of *C. elegans* in detail. Multiple nuclei can occasionally become transformed in a single injection. Additional studies of the metabolism and uptake of injected DNA in *C. elegans* gonads are of both technical and developmental import: for example, perhaps DNA injected into the cytoplasmic syncytium is packaged into nuclei as in *X. laevis* oocytes (Forbes et al. 1983).

After injection into *C. elegans,* foreign DNA forms high molecular weight arrays. Such arrays could be kinetically favored upon introduction of exogenous DNA or the concatenates could be a prerequisite structure for subsequent replication or segregation. In one injection, exogenous DNA was found in monomeric, circular form (data not shown). Single worms carrying the circular DNA could not be isolated. The circular molecules may have been very unstable.

In many DNA transformation systems, the introduction of linear DNA has important consequences. In yeast, the ends of linear molecules are highly recombinogenic; in mammalian systems, linear DNA increases the frequency of transformation; in the transformation of the sea urchin, linear DNA is absolutely required (Katula et al., this volume). In *C. elegans,* the only observable effect of linear DNA upon injection is to alter the mechanism of formation of the high molecular weight array. The linear DNA shows a facility for end-to-end ligation. This ligation of linear molecules could account for the formation of the minor species of injected sequences we observed. If DNA was randomly cleaved by endonucleases after injection, the ligation of the linear molecules would create random duplications and deletions.

The ligation of linear DNA in eukaryotes is a longstanding observation. For example, the recently broken ends of maize chromosomes will fuse with one another (McClintock 1938). Surprisingly, McClintock found that such broken chromosomes could be "healed" if they were included in diploid zygote tissue (McClintock 1941). Perhaps telomeres are added to injected DNA in *C. elegans,* stabilizing the extrachromosomal arrays.

In fact, the extrachromosomal high molecular weight arrays are surprisingly stable in *C. elegans.* An adult hermaphrodite contains approximately 1000 germline nuclei (Hirsh et al. 1976), generated from two embryonic precursors (Sulston et al. 1983) in a minimum of 14 mitotic cell divisions (five to generate Z2 and Z3, another nine assuming simple proliferative division). Thus a stability of 50% per worm generation corresponds to a maximum 5% rate of loss per cell division. In mammalian cells, unintegrated arrays are lost in 3% of the cells per cell division (Capecchi 1980; Scangos and Ruddle 1981). Extrachromosomal concatamers have also been observed in *Schizosaccharomyces pombe*; they are lost at a rate of 5–13% per division (Sakaguchi and Yamamoto 1982). Clarke and Carbon (1983) have deleted the centromere of chromosome III of *Saccharomyces cerevisiae*. Their data indicate that the acentric chromosome is lost at a rate of 10–15% per cell division. Returning to *C. elegans* for one final comparison, chromosomal fragments known as free duplications also show stabilities of approximately 50% per worm generation (Herman et al. 1979). Yet the formation of high molecular weight arrays with similar stabilities does not require any specific worm sequences. Indeed, the plasmids that successfully transformed *C. elegans* share only 2.9 kbp of pBR322 sequences. We are currently exploring the factors that may contribute to the stability of the extrachromosomal concatamers.

We have shown that the foreign sequences in these arrays can be expressed. If such expression is under proper spatial and temporal regulation, we can contemplate the use of these transformation techniques to study genes of developmental import. The extrachromosomal location of the foreign DNA precludes concerns of position effects that could be observed upon integration into random genomic locations. However, the presence of rearranged plasmid sequences in the high molecular weight arrays is a serious drawback to studies of sequences that may regulate developmental expression. It would be difficult to establish that expression is due to the original injected sequence rather than a rare duplication or deletion. Perhaps properly packaged DNA (condensed with histones) or suitably engineered vectors (such as linear molecules capped with functional telomeres) would not undergo these vexatious rearrangements. Difficulties might also be encountered due to the high copy number and the occasional loss of the injected sequences. Problems with gene dosage and overexpression might be overcome by simply diluting the injected sequences with an inert carrier DNA. Mosaicism is not likely to prevent proper expression. We have shown that one-half of the progeny of a transformant still carry some exogenous sequences and approximately one-half of these transformed animals contain foreign DNA in every cell. Thus, we would expect to be able to observe expression of genes that are required in specific cells or tissues with a penetrance of at least 25%. However, although mosaics are a bane to the molecular biologist concerned with gene expression, they are a boon to the developmental biologist. The analysis of mosaics in *C. elegans* allowed Herman (1984) to address the anatomical location of gene action. Is a cell or tissue-specific defect due to a lesion intrinsic to those cells, or is the defect manifested through interactions with other cells? Generating mosaics by DNA transformation would augment such studies.

ACKNOWLEDGMENTS

The authors would like to thank Alexandra Korte for her superb technical assistance and Dr. Victor Ambros for his critical reading of the manuscript. D.T.S. was a fellow of the Jane Coffin Childs Fund for Medical Research. This work was performed under the auspices of NIH Public Health Service Grant GM 19851.

REFERENCES

Ambros, Y. R. and H. R. Horvitz. 1984. Heterochronic mutants of the nematode, *Caenorhabditis elegans. Science* **226**: 409.

Brenner, S. 1974. The genetics of *Caenorhabditis elegans. Genetics* **77**: 71.

Capecchi, M. R. 1980. High efficiency transformation by direct microinjection of DNA into cultured mammalian cells. *Cell* **22**: 479.

Clarke, L. and J. Carbon. 1983. Genomic substitutions of centromeres in *Saccharomyces cerevisiae. Nature* **305**: 23.

Files, J. G., S. Carr, and D. Hirsh. 1983. Actin gene family of *Caenorhabditis elegans. J. Mol. Biol.* **164**: 355.

Forbes, D. J., M. W. Kirschner, and J. W. Newport. 1983. Spontaneous formation of nucleus-like structures around bacteriophage DNA microinjected into *Xenopus* eggs. *Cell* **34**: 13.

Greenwald, I. S., P. W. Sternberg, and H. R. Horvitz. 1983. The *lin-12* locus specifies cell fates in *Caenorhabditis elegans. Cell* **34**: 435.

Herman, R. K. 1984. Analysis of genetic mosaics of the nematode *Caenorhabditis elegans. Genetics* **108**: 165.

Herman, R. K., J. E. Madl, and C. K. Kari. 1979. Duplications in *Caenorhabditis elegans. Genetics* **92**: 419.

Hirsh, D., D. Oppenheim, and M. Klass. 1976. Development of the reproductive system of *Caenorhabditis elegans. Dev. Biol.* **49**: 200.

Kimble, J. and D. Hirsh. 1979. The postembryonic cell lineages of the hermaphrodite and male gonads in *Caenorhabditis elegans. Dev. Biol.* **70**: 396.

Kimble, J., J. Hodgkin, T. Smith, and J. Smith. 1982. Suppression of an amber mutation by microinjection of suppressor tRNA in *C. elegans. Nature* **299**: 456.

Kramer, J. M., G. N. Cox, and D. Hirsh. 1982. Comparisons of the complete sequences of two collagen genes from *Caenorhabditis elegans. Cell* **30**: 599.

———. 1985. Expression of the *C. elegans* collagen genes *col-1* and *col-2* is developmentally regulated. *J. Biol. Chem.* **260**: 1945.

MacLeod, A. R., J. Karn, and S. Brenner. 1981. Molecular analysis of the *unc-54* myosin heavy-chain gene of *C. elegans. Nature* **290**: 386.

McClintock, B. 1938. The fusion of broken ends of sister half-chromatids following chromosome breakage at meiotic anaphases. *Mo. Agric. Exp. Stn. Res. Bull.* **290**: 1.

———. 1941. The stability of broken ends of chromosomes in *Zea mays. Genetics* **26**: 234.

Orr-Weaver, T. L., J. W. Szostak, and R. J. Rothstein. 1981. Yeast transformation: A model system for the study of recombination. *Proc. Natl. Acad. Sci. U.S.A.* **78**: 6354.

Sakaguchi, J. and M. Yamamoto. 1982. Cloned *ural* locus of *Schizosaccharomycces pombe* propagates autonomously in this yeast assuming a polymeric form. *Proc. Natl. Acad. Sci. U.S.A.* **19**: 7819.

Scangos, G. and F. H. Ruddle. 1981. Mechanisms and applications of DNA-mediated gene transfer in mammalian cells–A review. *Gene* **14**: 1.

Southern, P. J. and P. Berg. 1982. Transformation of mammalian cells to antibiotic resistance with a bacterial gene under control of the SV40 early region promoter. *J. Mol. Appl. Genet.* **1**: 327.

Sulston, J. E. and H. R. Horvitz. 1977. Post-embryonic cell lineages of the nematode, *Caenorhabditis elegans. Dev. Biol.* **56**: 110.

Sulston, J. E., E. Schierenberg, J. G. White, and J. N. Thompson. 1983. The embryonic cell lineage of the nematode *Caenorhabditis elegans. Dev. Biol.* **100**: 64.

Concluding Remarks

RUDOLF JAENISCH
Whitehead Institute and Department of Biology
Massachusetts Institute of Technology
Cambridge, Massachusetts 02142

In recent years, considerable advances have been achieved in our ability to manipulate the mammalian embryo. This volume has highlighted several major experimental approaches that are likely to have significant impact on our understanding of molecular mechanisms of gene control in embryonic development as well as in differentiated cells of the adult. In the following paper, I will restrict my remarks to studies with mammalian embryos and focus on the major conclusions from nuclear transplant experiments, embryonic stem cell systems, and transgenic mice as used for insertional mutagenesis and control of gene expression.

Nuclear transfer experiments demonstrate that normal mammalian development critically depends not only on the presence of a diploid set of parental genes but also on the functional state of the genome. Zygotes that contain two maternal pronuclei (gynogenones) or two paternal pronuclei (androgenones) do not develop to term but die soon after implantation. The maternal genome seems to be sufficient for development to the blastocyst stage and to be essential for development of the embryo proper, but fails to support the proliferation of extraembryonic tissues. By contrast, the paternal genome is essential for trophoblast development. It appears that the paternal and maternal genomes are not identical and that the embryo is hemizygous for at least some genes whose expression is needed in development. Functional hemizygosity is likely to be a consequence of differential imprinting of the genome during gametogenesis. During cleavage the functional stage of the nucleus changes quickly as diploid nuclei from later stage embryos are not able to support development when transferred to the zygote. Understanding the structural changes imprinted on the mammalian genome during gametogenesis and early cleavage appears to be of major importance for our comprehension of gene control in development.

Embryonic stem cells can be derived directly from explanted embryos and propagated in vitro. When introduced into developing embryos, these cells will form chimeras and, most significantly, will colonize the germline. In principle, this system would seem suitable to introduce genes into cultured cells and to derive a mouse strain with a phenotype which was selected for in vitro. First attempts to derive chimeras from stem cells which were transfected with recombinant DNA or infected with retrovirus vectors have been successful. Germline contribution of manipulated stem cells has not been achieved yet and it remains an open question whether a diploid karyotype can be preserved after long-term culture and selection of the stem cells.

Insertional mutagenesis in mice can result from infection of embryos with retroviruses or from microinjection of recombinant DNA into zygotes. Because retroviruses insert as single copies into the host genome, a mutated gene can be easily cloned by using the provirus as molecular tag. This has been done for the collagen I gene which was mutated by a retrovirus insertion resulting in an embryonic lethal phenotype. The resulting mutant mouse strain can be used to study the role of collagen in early morphogenesis. In certain hybrid mice endogenous retroviruses appear to be unstable, giving rise to new germline integrations. It remains to be seen whether this system can be exploited to generate spontaneous virus-induced mutations. While it has been shown that the virus-induced mutation of the collagen gene involves a change of the chromatin conformation and of the methylation pattern, the molecular anatomy of genes that have been mutated by recombinant DNA has not yet been elucidated. The remarkably high frequency of mutations induced by DNA injection, as opposed to the low frequency of virus-induced mutations, may suggest, however, that introduction of DNA could induce major rearrangements or deletions that cause mutations in the host cell. It is likely that the use of insertional mutagenesis by retroviruses or by DNA injection has great potential for dissecting molecular mechanisms of mammalian development.

The expression of genes introduced into the very first transgenic mice was low and often occurred in nonappropriate tissues. The chromosomal position of the inserted genes appeared to have influenced the pattern of expression in an unpredictable manner. More recently derived transgenic mice, however, often activate the inserted gene in a highly tissue-specific manner, independently of the chromosomal position, as has been demonstrated for immunoglobulin genes and for the elastase gene. It appears that bacterial sequences interfere with normal gene activation as a high fraction of transgenics expressed the globin gene or the α fetoprotein gene in the appropriate tissue when the vector sequences had been removed prior to microinjection. The technique of generating transgenic mice has become so efficient as to allow deletion mapping of *cis*-acting DNA elements involved in developmental programming of gene activation. Furthermore, tissue-specific promoters can be used to target expression of oncogenes to specific tissues. SV40 T-Ag under control of the elastase or the insulin promoter induces specifically tumors of exocrin, or endocrin pancreas cells, respectively, when introduced into transgenic mice. Likewise, the *myc* gene under the hormone-stimulated mouse mammary tumor virus promoter induces carcinomas in mammary glands of lactating females. Thus it may be possible to study the biological effect of expressing a given gene in any cell type by placing the gene under the control of the tissue-specific promoter of choice and introducing it into transgenic mice.

In summary, this volume demonstrates that the tools to manipulate mammalian embryos and to derive transgenic animals have been refined to a level of sophistication that allows examination of major unresolved questions of the biology of mammals. These include questions of development, of immune surveillance and of oncogenesis, to name only a few. It is likely that this new experimental approach will yield a wealth of new information which hardly could have been anticipated a few years ago.

CORPORATE SPONSORS OF COLD SPRING HARBOR LABORATORY

Agrigenetics Corporation
Biogen, Inc.
CPC International, Inc.
E.I. du Pont de Nemours & Company
Genentech, Inc.
Genetics Institute
Hoffmann-La Roche Inc.
Johnson & Johnson
Eli Lilly & Company
Molecular Genetics, Inc.
Monsanto Company
Pall Corporation
Pfizer, Inc.
Schering Corporation
Upjohn Company

CORE SUPPORTERS OF THE BANBURY CENTER

The Bristol-Myers Fund, Inc.
The Dow Chemical Company
Exxon Corporation
Grace Foundation Inc.
International Business Machines Corporation
Phillips Chemical Company
Texaco Philanthropic Foundation Inc.

Name Index

Subject Index